CREATIVE
CHINESE
HOME
DISHES

創意
家常菜

出版序
PUBLICATION ORDER

時光飛逝，距母親離開我們已經快 20 年了，在這些年當中，弟弟跟我將母親的一生著作，陸續重新整理再出版，期間有些照片因保留不善無法使用，因此由我自己將菜重新做過，重新拍照出版；但是也有像《中國名菜精選》裡的菜色，因為有許多菜的難度實在太高，不敢輕易嘗試重作，而且原版的照片畫質清晰，所以保留原照片，只在版面編排上做了一些變化，使它更有中國風味。有時想想，當年母親出版這些食譜時的毅力與決心，真是我們所難及的。

《創意家常菜》，第一次出版是在 1994 年，選的菜大部份是經過媽媽絞盡腦汁，在做法跟口味上求新求變所創作出來的，雖然在那個時代是有新意、卻略有難度，但是以現在的選取食材的方便性來看，應該仍是有變化，但是難度降低，適合家庭主婦在家嘗試做給家人享用，故仍保持原來的書名——《創意家常菜》。

去年公司搬遷，有幸在堆積如山的底片中，找到了這本書保存完整的底片，再加以重新編排，終於有機會使本書得以重新跟讀者見面，跟大家分享母親的創意，希望大家會喜歡。

程安琪

作者序
PREFACE

　　時光飛逝如梭，一轉眼，從事電視烹飪教學已達卅年餘，為求與過去四千五百多次節目內容之不重複，筆者絞盡腦汁，不停地在口味上、做法上求新求變。而在選料上則一本以往一貫作風，以容易購得、經濟實惠為原則，以便能深入每個家庭。而各類菜餚之收集，包括禽、畜、海鮮、蔬果、素、魚等等，既可供家常之需亦有多項適宜宴客參考。（本書之內容係採自傅培梅時間自81年～83年示範內容之精選。）

　　中菜變化之多，可謂包羅萬象：一菜一格，百菜百味，浩瀚而無涯。只要膽大心細，多加練習，小心地避免火候、刀工、熱油之傷害，而在口味的調理上不必墨守成規，以合家人之鹹淡適口為依歸，如此則主中饋並非難事也。

　　烹調雖辛苦，但是除了做好菜來滿足家人的胃以外，還有什麼更能貼切的表達我們的愛和耐心呢？

How time flies. It's been over 30 years since I started to demonstrate my cooking program in TV company. In order not to repeat my 4500 times recipes, I squeeze my brain to try to invent new dishes technically and flavor-wise. To choose the ingredients, I pick up those that are cheaper and easy to get, so that it will be more convenient for housewives.

The recipes includes poultry, meat, fish, sea food and Vegetables, both good for daily meal and Banquet treat.

The varieties of Chinese cookings are countless, you have to cook carefully with plenty of guts, avoid hurting from knife or hot oil, and to season according each family's different taste. Though cooking is hard and easy to get hurt. Except cooking, what else can exactly explain the love feelings and patience so deeply?

CONTENTS ·

鳳翼雙味

Two Ways Chicken Wings

01　雞翅洗淨後，翅尖切除不用，將翅膀及翅根分割開，翅根部分將皮往下推成為球狀。用拌雞料拌勻，全部雞翅放入熱油中；炸黃撈出。

02　將炸過之翅膀、翅根加剩下之拌雞料和紅燒料一起紅燒 20 分鐘；將翅根球撿出。

03　乾粉絲用熱油炸泡，瀝乾，排入盤中，紅燒翅膀盛在粉絲上。

04　翅根球裹上酥炸糊，炸黃後排在周圍便可上桌，不用炸粉絲亦可用生菜切絲墊底。

Tips

乾粉絲可排在模型中炸成碗狀較為美觀。

INGREDIENTS •----------

10 Chicken wings, 5g. Mung bean thread.

SEASONINGS

 3tbsp. Soy sauce, 1tbsp. Wine.

❷ 2 Green onion, 2 Ginger slices, ½ Star anise, 1tbsp. Sugar, 2cups Water.

❸ 3tbsp. Flour, 2tbsp. Cornstarch, 1tsp. Baking powder, ⅔ cup Water.

STEP BY STEP •----------

01　Clean the chicken wings, cut off the ends from wings, then cut each wings into 2 parts, pull the meat to the end of the bone, like a lollipop. Marinate all with ❶, deep fry in hot oil to golden brown, drain.

02　Cook chicken with remaining ❶ and ❷ for 20 minutes.

03　Take out chicken balls (lollipop), coat with ❸ and deep fry to golden brown, arrange around the plate.

04　Deep fry mung bean thread in hot oil, drain and put on center of a plate, then put the middle part of wings on mung bean thread. Serve. You may use the lettuce shreds to replace the mung bean thread.

You may arrange the dried mung bean thread in a mold to keep the shape well after deep-fried.

五彩碎米雞

Minced Chicken with Popped Rice

01　雞胸肉切小丁後用醃雞料拌勻，醃 10 分鐘以上。

02　鍋巴用熱油炸泡，待冷後切碎放在盤中（可買炸好之鍋巴）。

03　炒鍋中將 1 杯油燒至七分熱後，放雞丁過油，待雞肉變白時撈出。然後將油倒出，僅留 1 湯匙在鍋中。

04　用餘油先爆香蔥、薑，放下香菇丁、筍丁一起炒熟，再放下雞丁和青椒丁、紅辣椒丁，淋下綜合調味料，大火拌炒均勻即可盛放在鍋巴上。

INGREDIENTS •

240g. Chicken breast, 4 pieces Pop rice cake, ⅓ cup Green pepper cubes, 2tbsp. Red chili cubes, ½ cup Black mushroom cubes, ½ cup Bamboo shoot cubes, 1tbsp. Green onion, chopped ,1tbsp. Ginger, chopped.

SEASONINGS

❶ ½tbsp. Soy sauce, ½tbsp. Cornstarch, 1tbsp. Water.

❷ 1tbsp. Soy sauce, ½tbsp. Wine, ½tbsp. Hot bean paste, 1tsp. Sugar, ½tsp. Salt, a few drops of Sesame oil, 1tsp. Cornstarch.

STEP BY STEP •

01　Cut chicken breast into small pieces, marinate with seasonings ❶ for 10 minutes.

02　Deep fry pop rice cake in hot oil. Let cool and crushed, arrange on a plate.

03　Heat 1cup of oil to deep fry chicken (about 150℃), when chicken turns white, drain.

04　Heat 1tbsp. of oil to fry green onion, ginger, mushroom, bamboo shoot, then add chicken, green & red pepper in, season with ❷ , stir over high heat thoroughly. Pour over popped rice.

醋溜雞花

Sweet & Sour Flower-shaped Chicken

01 雞胸（或腿）去骨，皮朝下平放在砧板上，直刀在雞肉上切細密交叉刀口，再分割成 1 寸大小塊狀，用醃雞料拌勻後醃 20 分鐘。

02 青椒、紅辣椒切小片；筍切成梳子片備用。

03 鍋中將 2 杯油燒至八分熱，放入雞肉泡熟、瀝出。然後另起油鍋，用 2 湯匙油爆香蔥、薑及蒜片，加入青椒、紅辣椒及筍片，大火炒一會兒後，加入雞肉和綜合調味料，拌炒均勻即可盛盤。

INGREDIENTS •

1 Chicken breast, 1 Green pepper, ½ Bamboo shoot, 2 Red chilis, 1tbsp. Green onion sections, 1tsp. Ginger slices, 1tbsp. Garlic slices

SEASONINGS

❶ 1tbsp. Egg white, ½tbsp. Wine, ½tsp. Salt, a pinch of pepper, 1tbsp. Cornstarch, 1tbsp. Oil.

❷ 1 ½tbsp. Soy sauce, 1 ½tbsp. Vinegar, 1tbsp. Wine, 1tsp. Sugar, ½tbsp. Cornstarch, 2tbsp. Water.

STEP BY STEP •

01 Remove bones from chicken breast, place on a chopping board, scare the meat part criss crossly, then cut into 1" cubes, marinate with seasoning ❶ for 20 minutes.

02 Cut green pepper and red chili into small pieces. Slice the bamboo shoot thinly and cut each slices into a comb shape.

03 Heat 2 cups of oil to 160℃, deep fry the chicken till done, drain. Heat another 2tbsp. of oil to stir fry green onion, ginger and garlic, add green & red chili, bamboo shoot in, stir fry over high heat, add chicken and seasonings ❷ , mix evenly. Serve.

蜜柑扣雞

Mold Chicken with Orange Sauce

使用材料

雞腿	2 支
柳丁	1 個
洋葱丁	½ 杯

調味料

番茄醬	½ 湯匙
酒	½ 湯匙
鹽	½ 茶匙

糖	2 茶匙
水	½ 杯
太白粉水	酌量

01 將雞腿去骨後，用醬油略拌。用熱油將表面炸黃，撈出切塊，鋪在湯碗中。

02 柳丁擠汁約 1 湯匙量；皮切細條。

03 用 1 湯匙油將洋蔥丁炒軟，加入番茄醬炒紅。另淋入酒、糖、鹽和水，煮滾後淋入①項的湯碗中，大火蒸半小時。

04 將蒸好的湯汁再慢慢倒入鍋中，加入柳丁汁及皮煮滾後，將汁中的渣質撈棄，淋下太白粉水勾芡，澆至扣在盤中的雞腿上即可。

INGREDIENTS •

2 Chicken legs, 1 Orange, ½ cup Onion, diced.

SEASONINGS

½tbsp. Ketchup, ½tbsp. Wine, 2tsp. Sugar, ½tsp. Salt, ½ cup Water, 2tsp. Cornstarch paste.

STEP BY STEP •

01 Remove bones from chicken legs, marinate the meat with soy sauce and deep fry in hot oil, cut into pieces and arrange in a bowl.

02 Squeeze juice out of the orange for later use. Peel the skin, use the very outer part (fine and thin as if you can see it through) , cut into fine shreds.

03 Heat 1tbsp. of oil to stir fry onion, add ketchup, wine, sugar, salt and water, bring to a boil, pour over chicken. Steam for ½ hour over high heat.

04 Pour the steamed chicken soup in a pan, add orange juice and peel, bring to a boil, drain off the peel, thicken with cornstarch paste, pour over chicken. Serve.

金杯松子雞

Chicken & Pine Nuts in Golden Cups

使用材料

青瓜（大黃瓜）	1 條
春捲皮	4 張
雞肉	6 兩
香菇丁、青椒丁	各 ½ 杯
紅辣椒段	2 湯匙
松子	2 湯匙
蔥小段、薑小片	酌量

拌雞料

醬油	½ 湯匙
太白粉	1 茶匙
水	½ 湯匙

綜合調味料

淡色醬油、酒、水	各 ½ 湯匙
鹽	¼ 茶匙
糖	½ 茶匙
胡椒粉、麻油	各少許
太白粉	1 茶匙

01 青瓜在最粗的地方切成兩半，用鋁箔紙包裹，並將底部壓平。

02 春捲皮（修剪成直徑約 20 公分圓形）平放到熱油中，並用青瓜棒由中間壓下，使春捲皮捲成杯子狀。待成形後，使青瓜與春捲皮分離，續將春捲皮炸成金黃色杯子，撈出瀝乾。

03 雞肉切小丁，用醃雞料拌醃 10 分鐘以上，用熱油將雞肉泡熟後瀝出。

04 另起油鍋，1 湯匙油爆香葱段、薑片，並加入雞丁等料同炒，淋下綜合調味料拌勻便熄火。撒下炸過之松子，分別盛入金杯中，裝盤上桌。

 Tips

‣ 松子先用糖水泡 20 分鐘，瀝乾，再用小火溫油炸黃，撈出後鋪在紙上待冷即可。

‣ 用半支青瓜，包裹上鋁箔紙以使春卷皮炸出一個杯子形狀，非常有趣，不妨試試。

INGREDIENTS •⋯⋯⋯⋯

1 Big cucumber, 4 pieces egg roll wrapper, 240g. Chicken breast, ½ cup Black mushroom cubes, ½ cup Green pepper cubes, 2 pieces Red chili, sectioned, 2tbsp. Pine nuts, 2 Green onion, sectioned, 3 pieces Ginger.

SEASONINGS

❶ ½tbsp. Soy sauce, 1tsp. Cornstarch, ½tbsp. Water.

❷ ½tbsp. Soy sauce, ½tbsp. Wine, ½tbsp. Water, ¼tsp. Salt, ½tsp. Sugar, a pinch of Pepper, a few drops of Sesame oil, 1tsp. Cornstarch.

STEP BY STEP •⋯⋯⋯⋯

01 Cut big cucumber into halves, wrap with a piece of aluminum foil, flat the bottom.

02 Heat 5cups of oil to very hot, place a piece of egg roll wrapper on surface of oil, press down immediately from center with the cucumber stick, to make the egg roll wrapper into a cup shape. When the cup is firmed, take out cucumber and drain the golden cup.

03 Dice chicken, marinate with seasoning ❶ for 10 minutes, deep fry in hot oil to done, drain.

04 Heat 1tbsp. of oil, stir fry green onion, ginger, add chicken, all ingredients and seasonings ❷ , mix well and turn off the heat. Add in fried pine nuts, pour into cups, serve.

 Tips

· Soak pine nuts in sugar water, drain & deep fry in warm oil over low heat until golden brown. Remove.

· Give this a try; it if very interesting. Use half a cucumber and wrap it in aluminum foil so that the deep-fried spring roll wrapper takes the shape almost like that of a cup.

椰香葡國雞

Baked Chicken Portuguese Style

使用材料

雞腿	2 支	酒	½ 湯匙
大馬鈴薯	1 個	鹽	1 茶匙
洋葱丁	⅔ 杯	糖	½ 茶匙
大蒜屑	1 湯匙		
葡萄乾（泡軟）	約 20 粒	**麵糊料**	
		油	4 湯匙
調味料		麵粉	4 湯匙
咖哩粉	1 ½ 湯匙	牛奶	¼ 杯
水	4 杯	椰漿	¼ 杯

01 雞腿斬成小塊，沾上麵粉炸黃；馬鈴薯切滾刀塊也炸黃。

02 用 2 湯匙油爆香大蒜屑及洋蔥丁，再加入咖哩粉炒香，注入清水等調味料，加入雞塊及馬鈴薯，小火煮 30 分鐘。然後撈出雞塊及馬鈴薯放入烤碗中，灑入葡萄乾。

03 另取一鍋，倒入油加熱，炒香麵粉，倒入過濾後的湯汁，慢慢攪勻並加入牛奶及椰漿，調成糊狀淋到烤碗中。放入烤箱中，以 220°C 烤至微乾黃即可。

INGREDIENTS •

2 Chicken legs, 1 Potato, ⅔ cup Onion, chopped, 1tbsp. Garlic, chopped, 20 pieces Raisin.

SEASONINGS

❶ 1 ½tbsp. Curry powder, 4cups Water, ½tbsp. Wine, 1tsp. Salt, ½tsp. Sugar.

❷ 4tbsp. Oil, 4tbsp. Flour, ¼cup Milk, ¼cup Coconut milk.

STEP BY STEP •

01 Cut chicken leg into small pieces, coat with flour and deep fry together with potato cubes to golden brown.

02 Heat 2tbsp. of oil to stir fry garlic, onion and curry powder, add in water and seasoning ❶, then add chicken & potato, cook over low heat for 30 minutes, strain the soup for later use, put chicken and potato in a baking ware, sprinkle raisins on top.

03 Heat 4tbsp. of oil to stir fry flour then add the curry soup in, stir and mix with milk and coconut milk to a paste, pour on top of chicken, bake for 10 minutes over 220°C until surface turns to golden brown. Serve.

蠔油去骨雞

Boneless Chicken with Oyster Sauce

使用材料			
雞腿	2 支	調味料	
西洋生菜	半棵	蠔油	1 湯匙
		太白粉水	1 茶匙
醃雞料			
醬油	2 湯匙		
酒	1 湯匙		
葱支、薑片	酌量		

01 雞腿剔除骨頭，用刀將白筋斬斷，放入醃雞料中醃半小時。

02 生菜切寬條，在滾水中川燙一下撈出（滾水中放油 ½ 湯匙，鹽 ½ 茶匙），瀝乾水分，鋪在盤中。

03 燒熱 2 杯油，雞皮面朝下，放入油中炸熟，趁熱切小塊，排放生菜上（或用 5 湯匙熱油將雞腿煎熟）。

04 剩餘的醃雞料加水半杯倒入鍋中煮滾，撈棄蔥薑，勾芡後加入蠔油拌勻，淋到雞肉上。

INGREDIENTS ·

2 Chicken legs, ½ head Lettuce.

SEASONINGS

❶ 2tbsp. Soy sauce, 1tbsp. Wine, 1stalk Green onion, a few Ginger slices.

❷ 1tbsp. Oyster sauce, 1tsp. Cornstarch paste.

STEP BY STEP ·

01 Remove bones from chicken, chop the meat side and cut through the tendon, marinate with ❶ for ½ hour.

02 Shred lettuce, blench in boiling water, (add ½tbsp. of oil & ½tsp. of salt in water previously) drain, arrange a plate.

03 Heat 2cups of oil to deep fry chicken until done, remove and cut into pieces, then and arrange on lettuce.

04 Add ½cup of water to cook with seasoning ❶ , bring to a boil, discard green onion and ginger, then mix with oyster sauce and thicken with cornstarch paste, pour over chicken.

金針豉汁雞

Chicken & Dried Lily Flower with Fermented Bean Sauce

使用材料

雞腿		2 支
香菇		4 朵
金針		⅓ 兩
嫩薑		15 小片
葱段		2 湯匙

調味料

豆豉		1 湯匙
醬油、酒、蠔油		各 ½ 湯匙

糖、鹽		各 ¼ 茶匙
胡椒粉、麻油		各少許
太白粉		1 茶匙

01 雞腿去骨，肉切成 1 寸大小塊狀；香菇泡軟切斜片；金針泡軟，擠乾水分，放在蒸盤中。

02 用油 1 湯匙炒香豆豉，並加入其他調味料，放下雞肉、嫩薑及香菇拌勻，平鋪放在金針菜上，大火蒸 15 分鐘。

03 另熱 1 ½ 湯匙油，放下蔥段煎香，趁熱淋到雞肉上，略加拌合，盛入盤中。

INGREDIENTS •

2 Chicken legs, 4 Black mushrooms, 15g. Dried lily flower, 15 pieces Young ginger slice, 2tbsp. Green onion sections.

SEASONINGS

❶ 1tbsp. Fermented black beans, ½tbsp. Soy sauce, ½tbsp. Wine, ½tbsp. Oyster sauce, ¼tsp. Salt, ¼tsp. Sugar, 1tsp. Cornstarch, a pinch of Pepper, a few drops of Sesame oil.

STEP BY STEP •

01 Remove bones from chicken, cut meat into 1" cubes. Slice soaked black mushrooms into halves. Soak dried lily flowers to soft, squeeze and put on a plate.

02 Heat 1tbsp. of oil to stir fry fermented black beans, add all seasonings in, mix with chicken, black mushroom and ginger. Arange on top of lily flower, steam over high heat for 15 minutes.

03 Heat 1 ½tbsp. of oil, fry green onion till fragrant, pour over chicken. Mix and serve.

三絲鴨捲

Tri-Color Duck Rolls

使用材料			拌鴨料	
鹽水鴨（或燒鴨）	¼ 隻		蛋	1 個
瘦豬肉	3 兩		醬油	½ 湯匙
筍	1 支		酒	1 茶匙
香菇	3 朵		鹽	½ 茶匙
韭黃	5 支		麻油	1 茶匙
嫩薑	少許		糖	¼ 茶匙
豆腐衣	3 張		胡椒粉	少許

01　鴨去骨取肉，切成 2 寸長之細條。豬肉煮熟也切成相同大小。嫩薑、煮熟的筍、泡軟的香菇，均切成絲狀；韭黃切成 1 寸長段。

02　①項各料盛在大碗中，加入拌鴨料拌勻，分別用豆腐衣包捲成長條筒狀（3 條）。

03　油 2 杯燒至七分熱，放下鴨肉捲，用中火慢炸至外層酥脆即可撈出，切段盛盤，可附上花椒鹽或甜酸醬以供沾食。

INGREDIENTS •

¼ Roast duck, 120g. Lean pork, 1 Bamboo shoot, 3 pieces Black mushroom, 5 stalks pieces yellow leek, Young ginger shreds, 3 pieces Bean curd sheet.

SEASONINGS

1 Egg, ½tbsp. Soy sauce, 1tsp. Wine, ½tsp. Salt, 1tsp. Sesame oil, ¼tsp. Sugar, a pinch of Pepper.

STEP BY STEP •

01　Remove bones from duck, cut into 2" long strips. Cut boiled pork into same size. Shred the cooked bamboo shoot. Shred the soaked black mushroom. Cut yellow leek into 1" sections.

02　Mix the above ingredients with seasonings, wrap and fold with dried bean curd sheet to a roll. Make 3 rolls.

03　Deep fry rolls in warm oil over medium heat until crispy, drain and cut into sections. Serve with sweet & sour sauce or brown pepper corn powder with salt.

醬爆櫻桃

Frog's Legs with Sweet Soybean Sauce

使用材料

田雞		1 斤
青椒		1 支
紅辣椒		2 支
蔥		2 支
大蒜片		15 片

醃料

醬油		½ 湯匙
太白粉		2 茶匙

綜合調味料

甜麵醬		1 湯匙
水、酒、醬油		各 ½ 湯匙
糖		1 茶匙
番茄醬		1 茶匙
麻油		少許

01 田雞僅取用雙腿，分割後用醃料醃 20 分鐘以上（放入冰箱中）。蔥、紅辣椒切斜段；青椒切斜塊。

02 燒熱 1 杯油至 8 分熟，將田雞腿過油炸熟瀝出。

03 另起油鍋，用 1 湯匙油爆香大蒜片及蔥段，再將綜合調味料倒入炒香，放下青椒、紅辣椒及田雞腿，大火快炒，拌炒均勻便可裝盤。

菜名之「櫻桃」係江浙人見田雞腿炸熟後腿肉漲圓而稱之。故本菜僅用田雞腿。家庭中烹調可將有肉的部分切塊來爆炒。亦可用雞肉來做「醬爆雞丁」。

INGREDIENTS •

600g. Frog's legs, 1 Green pepper, 2 pieces Red chilies, 2 stalks Green onion, 15 pieces Garlic slice.

SEASONINGS

❶ ½tbsp. Soy sauce, 2tsp. Cornstarch.

❷ 1tbsp. Sweet soybean paste, ½tbsp. Wine, ½tbsp. Water, ½tbsp. Soy sauce, 1tsp. Sugar, 1tsp. Ketchup, a few drops of Sesame oil.

STEP BY STEP •

01 Cut each frog's leg into 2 parts, marinate with ❶ for 20 minutes. Cut green onion & red chili into sections, cut green pepper to pieces.

02 Heat 1cup of oil to 160°C, fry frog's legs to done, drain.

03 Heat 1tbsp. of oil to fry garlic and green onion, add seasonings ❷ , bring to a boil. Put green pepper, red chilies and frog s legs, stir fry over high heat, mix evenly. Serve.

Shanghai cuisine call this dish "Cherry with Sweet Soybean Sauce" because the frog's legs shrink after fried till done and look like cherries. You may use chicken to instead of frog's legs.

紅米醬肉

Pork with Red Rice Sauce

使用材料	五花肉	1 ½ 斤	酒	½ 杯
	紅米	½ 杯	冰糖	½ 杯
	葱	4 支		
	薑	2 片		
	八角	1 顆		

調味料

淡色醬油 ⋯⋯⋯⋯ 4 湯匙

01 五花肉整塊連皮切成 3 長條，用開水燙過，放入墊有蔥段、薑片之砂鍋或湯鍋中，加入滾水 3 杯及八角，小火煮半小時以上至肉已半爛為止。

02 紅米用 2 杯水泡軟後，用紗布袋包紮，將紅米汁擠入鍋中，紗布包亦放入，再加調味料。先用大火煮滾約 5 分鐘，繼續用小火燒煮 1 小時半以上至肉已十分軟爛。

03 食前取出，切成約 2 公分寬度，裝盤淋上肉汁即可上桌。

Tips

這道菜因烹調較費時，因此一次可做多些，燒好浸泡在湯汁中，食前另行加熱。

INGREDIENTS •

900g. Pork (bacon part), ½ cup Red rice, 4 stalks Green onion, 2 pieces Ginger slice, 1 Star anise.

SEASONINGS

❶4tbsp. Light colored soy sauce, ½ cup Wine, ½cup Rock sugar.

STEP BY STEP •

01 Cut pork belly into 2" thick slices about 10" long. Make 3 pieces. Blench in boiling water, drain, put into a casserole with green onion, ginger. Add 3cups of boiling water & star anise, cook over low heat for ½ hour.

02 Soak red rice with 2cups of water until soft, wrap in a cloth, tie up and squeeze the juice out, pour this juice and rice package to the casserole. Add seasonings and cook over high heat for 5 minutes, then turn to low heat, cook for another 1 ½ hours.

03 When serve, cut into 2cm cubes and pour over some sauce.

Tips

This dish needs long time to cook, you may cook a large portion and serve separately.

銀紙松子肉

Pine Nuts Pork in Silver Sticks

01 豬肉切成細條，蔥、薑拍碎加水 2 湯匙，擠出蔥薑水拌入肉中，再加入調味料拌攪，最後加入松子。

02 鋁箔紙刷上少許麻油，放入約 1 ½ 湯匙的①項料，包捲成小棒狀。

03 投入熱油中炸約 2 分鐘即可，撈出，瀝乾油，放入盤中，另以炸蝦片圍飾。

INGREDIENTS

300g. Pork(should part), 2tbsp. Pine nuts, 8 pieces Aluminum foil (8cm x 8cm). Dried shrimp chips, 2 stalks Green onion, 2 pieces Ginger.

SEASONINGS

1tbsp. Soy sauce, ½tsp. Sugar, ¼tsp. Salt, a pinch of Pepper, a pinch of Five-spicy powder.

STEP BY STEP

01 Cut pork into fine shreds. Pat the green onion & ginger, soak with 2tbsp. water, squeeze out the juice to mix with pork & seasonings, then mix with pine nuts.

02 Brush some sesame oil on the aluminum foil, put in 1 ½tbsp. pork mixture. Wrap & fold into a roll.

03 Deep fry in hot oil for 2 minutes. Put on a plate. Serve with deep fried shrimp chips.

什錦燴響鈴

Deep Fried Bells with Assorted Ingredients

使用材料

豬肉或雞肉片、魷魚、花枝、蝦仁、
豬肚、海參等⋯⋯⋯⋯⋯⋯⋯各酌量
筍片、胡蘿蔔片、豌豆夾、玉米筍、
香菇、草菇等蔬菜料⋯⋯⋯⋯⋯隨意
蔥、薑⋯⋯⋯⋯⋯⋯⋯⋯⋯各少許
絞肉⋯⋯⋯⋯⋯⋯⋯⋯⋯⋯⋯3兩
餛飩皮⋯⋯⋯⋯⋯⋯⋯⋯⋯12張

調味料

高湯⋯⋯⋯⋯⋯⋯⋯⋯⋯⋯⋯3杯
醬油⋯⋯⋯⋯⋯⋯⋯⋯⋯⋯½湯匙
鹽⋯⋯⋯⋯⋯⋯⋯⋯⋯⋯⋯1茶匙
胡椒粉⋯⋯⋯⋯⋯⋯⋯⋯⋯⋯少許
太白粉水⋯⋯⋯⋯⋯⋯⋯⋯1湯匙

01 絞肉加少許醬油及鹽，調成絞肉餡，用餛飩皮包成小餛飩。

02 將選用的各種材料分別切成片狀，用滾水川燙一下撈出。

03 加入 2 湯匙油，起油鍋爆香蔥、薑後，淋入高湯煮滾，再將各種材料依硬度先後下鍋，調味後勾芡，裝在大碗中保溫。

04 熱油將小餛飩炸熟成金黃色，裝在大碗中。淋入③項，使其有響聲，快速分食。

Tips

什錦料不拘種類，可隨意挑選調配。

INGREDIENTS •

Choose any of Pork, Chicken, squid, Cuttlefish, Shrimp, Pork stomach, Sea cucumber, Bamboo shoot, Carrot, Snow peas, Black mushroom, Straw mushroom or Baby corn, some Green onion, some Ginger, 12 pieces Won-Ton wrapper, 120g. Minced pork.

SEASONINGS

❶ 3cups Soup stock, ½tbsp. Soy sauce, 1tsp. Salt, a pinch of Pepper, 1tbsp. Cornstarch paste.

STEP BY STEP •

01 Mix minced pork with some soy sauce & salt, stir till very sticky, wrap in Won-Ton wrapper.

02 Cut every ingredient into slices. Blench in boiling water.

03 Heat 2tbsp. of oil to stir fry green onion & ginger, add in soup stock. When boiling, add all ingredients (putting in a sequence according to the texture of each ingredient), seasoned and thickened.

04 Deep fry Won-Ton to golden brown, put in a deep plate. Pour over the thickened assorted ingredients while serving.

Tips

You may add no matter what kind of ingredients which you prefer.

京都子排

Spareribs with King-Tu Sauce

使用材料		
豬小排	12 兩	
洋葱絲	1 杯	
醃肉料		
醬油	1 湯匙	
太白粉、麵粉	1 ½ 湯匙	
水	2 湯匙	
小蘇打粉	¼ 茶匙	

綜合調味料	
番茄醬、辣醬油、A I 牛排醬	各 1 湯匙
清水	2 湯匙
糖	½ 湯匙
麻油	½ 茶匙

01　將豬小排切成 2 寸寬（骨頭粗的由中間劈開），用混合均勻的醃肉料醃 1 小時以上。

02　用少量油炒熟洋蔥絲，加鹽調味，盛入盤內墊底。

03　炸油燒熱，放入豬小排，大火炸 2～3 分鐘，見外皮酥脆即可撈起。

04　另用 1 湯匙油炒煮綜合調味料，煮滾後將排骨下鍋拌合，即可盛放在洋蔥上。

INGREDIENTS •

450g. Spareribs, 1cup Onion shreds.

SEASONINGS

❶ 1tbsp. Soy sauce, 1 ½tbsp. Cornstarch, 1 ½tbsp. Flour, 2tbsp. Water, ¼tsp. Baking soda.

❷ 1tbsp. Ketchup, 1tbsp. A1 Sauce, 1tbsp. Worcester sauce, 2tbsp. Water, ½tbsp. Sugar, ½tsp. Sesame oil.

STEP BY STEP •

01　Cut spareribs into 2" sections, cut and split (from the bone) into 2 parts, marinate with seasonings ❶ for 1 hour.

02　Stir fry onion with 2tbsp. hot oil, season with salt. Remove to a plate.

03　Deep fry spareribs over high heat for 2~3 minutes till golden brown.

04　Heat 1tbsp. of oil to stir fry seasonings ❷ . When boiling, mix with fried spareribs. Put on fried onion.

爆羊肉雙味

Quick Stir Fried Lamb with Double Flavors

Dish. 1 沙茶羊肉片

使用材料

羊肉片	6 兩	調味料	
大蒜屑	1 湯匙	沙茶醬 ——— 1 ½ 湯匙	糖 ——— ¼ 茶匙
青菜	隨意	醬油、酒 —— 各 ½ 湯匙	

步驟說明

01　小碗中先將調味料調勻備用。

02　起油鍋炒熟青菜，瀝乾湯汁，排放在盤中間。

03　用 2 湯匙油爆香大蒜屑，放下羊肉片大火爆炒至熟，淋入調味料拌勻，盛在青菜的一邊。

Dish. 2 葱爆羊肉

使用材料

羊肉片	6 兩	醃肉料		調味料	
大蒜片	1 湯匙	醬油	1 湯匙	醬油	½ 湯匙
葱絲	1 杯	酒	½ 湯匙	麻油	1 茶匙
香菜	半杯	花椒粉	少許	醋	2 茶匙
		油	½ 湯匙		

步驟說明

01　羊肉用醃肉料醃 10 分鐘以上。

02　燒熱 2 湯匙油爆香大蒜片，馬上放下羊肉片，大火拌炒，見肉片轉白，放下葱絲續炒。淋下調味料快速炒勻便可盛出。也可在最後加入香菜段少許一起拌合。

DISH.1

Lamb with Sa-Cha Sauce

INGREDIENTS •

240g. Lamb slices, 1tbsp. Garlic, chopped, Green vegetables.

SEASONINGS

1 ½tbsp. Sa-Cha Sauce, ½tbsp. Soy sauce, ½tbsp. Wine, ¼tsp. Sugar.

STEP BY STEP •

01　Mix seasonings in a bowl.

02　Stir fry the vegetable and put on center of a plate.

03　Heat 2tbsp. of oil, stir fry garlic, add lamb and stir fry quickly over high heat until done, mix with seasonings and place on one side of the vegetables.

DISH.2

Lamb with Green Onions

INGREDIENTS •

240g. Lamb slices, 1tbsp. Garlic, sliced, 1cup Green onion shreds, ½ cup Chinese parsley.

SEASONINGS

❶ 1tbsp. Soy sauce, ½tbsp. Wine, ½tbsp. Oil, a pinch of Brown pepper corn powder.

❷ ½tbsp. Soy sauce, 1tsp. Sesame oil, 2tsp. Vinegar.

STEP BY STEP •

01　Marinate lamb with seasoning ❶ for 10 minutes.

02　Fry garlic in 2tbsp. of heated oil, add lamb and stir fry over high heat. When meat turns white, add green onion & seasonings ❷ . Mix well and serve. You may add Chinese parsley to fry at last.

錦繡牛肉捲

Beef Rolls

使用材料

大薄片嫩牛肉	6 片
白蘿蔔絲、胡蘿蔔絲、小黃瓜絲、	
生菜絲、香菜絲	各酌量

沾料

醬油	1 湯匙
麻油	½ 湯匙
高湯	3 湯匙
炒過的芝麻	1 茶匙

01　將牛肉片攤開，撒下少許鹽、胡椒粉，逐片用少許油煎熟。

02　白蘿蔔絲、胡蘿蔔絲、小黃瓜絲、香菜絲及生菜絲先混合後，包在牛肉片中，分別捲成筒狀，放在盤中。

03　將調好的沾料淋到牛肉捲上，並灑上炒過的芝麻即可。

INGREDIENTS •

6 Beef tenderloin slices, Turnip shreds, Carrot shreds, Cucumber shreds, Lettuce shreds, Chinese parsley shreds (each a little).

DIPPING SAUCE

1tbsp. Soy sauce, ½tbsp. Sesame oil, 3tbsp. Soup stock, 1tsp. Fried sesame seeds.

STEP BY STEP •

01　Spread the beef, sprinkle some salt & pepper, fry both sides with a little of oil.

02　Mix turnip, carrot, cucumber, lettuce and Chinese parsley, arrange on the beef, roll into a cylinder, put on a plate.

03　Pour the mixed dipping sauce over beef rolls, sprinkle some stir fried sesame seeds, then serve.

沙茶牛仔骨

Beef Spareribs with Sa-Cha Sauce

使用材料			調味料	
	牛小排	4 片	沙茶醬	1 湯匙
	葱段	1 湯匙	醬油	½ 湯匙
	薑片	4 小片	酒	½ 湯匙
			水	1 湯匙
	醃料			
	醬油	1 湯匙		
	太白粉	1 茶匙		
	水	2 茶匙		

01　將牛小排按骨縫分割成小塊，全部放入大碗中，用醃料拌醃。

02　用熱油 5 ～ 6 湯匙將牛小排大火煎熟，瀝出（或用多量油炸熟）。

03　用 1 湯匙油爆香蔥段及薑片，放入牛小排，再加入調味料，大火炒煮均勻，即可盛盤。

INGREDIENTS

4 pieces Beef spareribs, 1tbsp. Green onion sections, 4 pieces Ginger slice.

SEASONINGS

❶ 1tbsp. Soy sauce, 1tsp. Cornstarch, 2tsp. Water.

❷ 1tbsp. Sa-Cha Sauce, ½tbsp. Soy sauce, ½tbsp.Wine, 1tbsp. Water.

STEP BY STEP

01　Cut beef spareribs into small sections, marinate with seasonings ❶.

02　Heat 5tbsp. of oil to fry both sides of spareribs over high heat, drain.

03　Heat 1tbsp. of oil to stir fry green onion & ginger, add beef spareribs & seasonings ❷ in, fry over high heat, mix evenly and serve.

中式牛肉派

Beef Pie Chinese Style

使用材料

絞牛肉	4 兩
洋蔥屑	⅓ 杯
芹菜屑	¼ 杯

調味料

醬油、酒	各 ½ 湯匙
鹽	¼ 茶匙
胡椒粉	少許

蛋麵糊

蛋	3 個
麵粉	3 湯匙
太白粉	1 湯匙
鹽	½ 茶匙
發泡粉	½ 茶匙

01 先用熱油 2 湯匙炒香洋蔥屑,再加入絞牛肉大火炒熟,加調味料後,撒下芹菜屑即盛出。

02 在大碗中將蛋打鬆後,將蛋麵糊調勻。

03 用平底鍋將 3 湯匙油燒熱,倒入一半的蛋麵糊料,中火煎至半熟時,放下牛肉料,平均鋪平,再將另一半之蛋麵糊料淋下,蓋住牛肉餡。用小火燜煎至熟。煎時分數次淋下一些油,且中途需翻面,以使兩面均呈金黃色。取出,切成尖角型,排入盤中。

INGREDIENTS •

150g. Minced beef, ⅓ cup Onion, chopped, ¼ cup Celery, chopped.

SEASONINGS

½tbsp. Soy sauce, ½tbsp. Wine, ¼tsp. Salt, a pinch of pepper.

BATTER

3 Eggs, 3tbsp. Flour, 1tbsp. Cornstarch, ½tsp. Salt, ½tsp. Baking powder.

STEP BY STEP •

01 Heat 2tbsp. of oil to fry onion, add beef and seasonings, fry over high heat, mix finally with celery.

02 Mix the batter.

03 Heat 3tbsp. of oil in a pan, pour half of batter, fry to half cooked , add beef, pour another half of batter to cover the beef, fry over low heat to form a cake. (Add some oil intermediately) when the color turns golden brown, remove and cut into pieces. Serve.

什蔬燴咖哩牛肉

Curry Beef with Assorted Vegetables

使用材料

牛肋條或牛腱	1 斤
洋葱屑	½ 杯
大蒜屑	1 湯匙
冷凍什錦蔬菜	1 ½ 杯

煮牛肉料

水	4 杯
葱	2 支
薑	2 片

八角	1 粒
酒	1 湯匙

調味料

咖哩粉	1 ½ 湯匙
鹽	1 茶匙
糖	½ 茶匙
太白粉水	酌量

01 牛肉用開水燙煮，滾後撈出，洗淨。煮牛肉料先煮開後，放下牛肉煮至八分爛（約 1 小時），取出，切成 1 寸四方大小。

02 用 2 湯匙油炒香洋葱屑、大蒜屑及咖哩粉，加入牛肉及湯汁（約 1 杯，不足可加水），再煮 5 分鐘。

03 放下冷凍蔬菜料再煮滾，加鹽、糖調味，勾芡後即可裝盤上桌，淋在白飯、熟麵條或通心粉上均可。

市面有售已調味之咖哩塊，可挑選喜愛口味，不需再調味及勾芡。

INGREDIENTS •

600g. Beef brisket, ½ cup Onion, chopped, 1tbsp. Garlic, chopped, 1 ½cups Frozen assorted vegetables.

SEASONINGS

❶ 4cups Water, 2 stalks Green onion, 2 Ginger slices, 1 Star anise,1tbsp. Wine.
❷ 1 ½tbsp. Curry powder, 1tsp. Salt, ½tsp Sugar, some Cornstarch paste.

STEP BY STEP •

01 Blanch beef in boiling water, drain and rinse. Boil seasoning ❶, add beef and cook for 1 hour. Remove and cut into 1" cubes.

02 Heat 2tbsp. of oil, stir fry onion, garlic & curry powder, add beef and soup, cook for 5 more minutes.

03 Add frozen vegetables, sugar and salt, thickened and serve with rice or noodles.

Tips

The curry cube can be used as a substitute for powder, don't season the beef again.

葱油淋漢堡

Chinese Hamburger with Green Onion

使用材料

全瘦絞牛肉	半斤	醬油	½ 湯匙
絞肥豬肉	2 兩	鹽	½ 茶匙
西洋菜	1 把	胡椒粉	少許
葱絲、嫩薑絲	各 ½ 杯	太白粉	1 湯匙

拌肉料

葱薑水	⅓杯
小蘇打	⅓茶匙
酒	½ 湯匙

01 絞牛肉與肥豬肉混合均勻，加入蔥薑水等拌肉料，用手抓拌，摔擲多次，使肉增加彈性（蔥薑水的比例為蔥 2 支、薑 3 片拍碎，泡入⅓杯水中）。

02 西洋菜切短，在開水中燙一下撈出，冷開水沖涼後，鋪在深盤中。

03 牛肉做成 6 個圓球，壓成扁圓形的漢堡排，放在西洋菜上，大火蒸 8 分鐘，再撒下蔥、薑絲，淋下燒得極熱之油，透出香氣即可。

INGREDIENTS ·

300g. Minced lean beef, 75g. Minced pork fat, a bunch Watercress, ½ cup Green onion shreds, ½ cup Ginger shreds.

SEASONINGS

⅓ tsp. baking soda, ½tbsp. Wine, ½tbsp. Soy sauce, ½tsp. Salt, ⅓ cup Ginger & green onion juice, 1tbsp. Cornstarch, a pinch of Pepper.

STEP BY STEP ·

01 Mix minced beef and pork fat, add seasonings and mix well, stir to sticky.(For green onion & ginger juice is crush 2 stalks green and 3 slices ginger, soak in 1/3 cup of water for 5 minutes.)

02 Cut watercress into sections 2" long, blanch and rinse cool, put on the plate.

03 Use beef to make 6 balls, press flat a little bit, arrange on the watercress, steam for 8 minutes. Sprinkle ginger & green onion shreds, splash heated oil over. Serve.

鍋燜牛里脊

Pan Stewed Beef

使用材料

牛里脊肉	6 兩
麵粉	2 湯匙
蛋黃	2 個

醃肉料

蔥、薑末	各少許
太白粉	1 湯匙
水	1 湯匙

調味料

高湯（或水）	1 杯
醬油	2 湯匙
酒	½ 湯匙
糖	1 茶匙
鹽	¼ 茶匙
胡椒粉	少許

01 將牛里脊肉橫片成⅓寸厚的大片，在肉面上劃切交叉刀口，用醃肉料醃10分鐘以上（可買切好之牛排再橫切成⅓寸厚片）。

02 牛肉先沾麵粉再沾滿蛋黃，用熱油煎黃。

03 用1湯匙油炒煮調味料，煮滾後將牛肉入鍋，小火煨煮約2分鐘。

04 夾出牛肉切寬條裝盤，淋上肉汁即可。

所謂鍋燜即是將食材沾過麵粉，再沾上蛋汁，然後用油煎黃的一種烹調技法，如鍋燜豆腐、鍋燜鮮貝。

INGREDIENTS •

240g. Beef tenderloin, 2tbsp. Flour, 2 Egg yolk.

SEASONINGS

❶ a little of Green onion & Ginger crushed, 1tbsp. Cornstarch, 1tbsp. Water.

❷ 1 cup Soup stock, 2tbsp. Soy sauce, ½tbsp. Wine, 1tsp. Sugar, ¼tsp. Salt, a pinch of Pepper.

STEP BY STEP •

01 Cut beef into ⅓ " thick pieces, score crisscross on surface, marinate with seasonings ❶ for 10 minutes.

02 Sprinkle flour on beef, dip in egg yolk, fry in hot oil till golden brown.

03 Heat 1tbsp. of oil to fry seasonings ❷ , when boiling, add beef in and cook over low heat for 2 minutes.

04 Take out beef and cut into 2" wide pieces, put on a plate, pour over the sauce. Serve.

Tips

So-called "potting" is a cooking technique that requires taking ingredients and dipping them in flour, then into beaten egg yolks, and finally pan frying them in oil until they turn yellow--producing dishes such as potted tofu or potted scallops.

生汁龍蝦球

Lobster Balls with Mayonnaise Sauce

使用材料		
龍蝦	1 隻
明蝦	3 隻
玉米粉	½ 杯
生菜絲	1 ½ 杯
醃蝦料	
蛋白	1 個
鹽	¼ 茶匙

拌蝦料	
美乃滋	½ 杯
芥末醬	1 茶匙
牛奶	1 湯匙

01 龍蝦及明蝦剝殼取肉後，切成 2 公分四方大小，用醃蝦料拌醃 10 分鐘以上。

02 生菜絲墊盤底，將龍蝦頭與尾蒸熟後刷上少許油，使其光亮後排在盤中。

03 蝦肉沾滿玉米粉後投入熱油中炸酥，撈出後放入調勻的拌蝦料碗中，抖動碗使蝦肉均沾上醬料，排入盤中央。

Tips

▸ 用明蝦是使蝦肉份量增加，也可以都用龍蝦或用明蝦來做。

▸ 廣東人稱美乃滋為生汁，這道菜源自香港師傅發想的「生汁芝麻蝦」，即目前我們所稱的鳳梨蝦球。

INGREDIENTS •

1 Lobster, 3 Prawns, ½ cup Cornstarch, 1 ½cups Lettuce leaf shreds.

SEASONINGS

❶ 1 Egg white, ¼tsp Salt.
❷ ½cup Mayonnaise, 1tsp. Mustard, 1tbsp. Milk.

STEP BY STEP •

01 Shell the lobster & prawns, cut the meat into 2" long sections, marinate with ❶ for 10 minutes.

02 Steam the head & tail of lobster and put on one side of a plate. Arrange lettuce leaves on center of the plate.

03 Coat lobster & prawns with cornstarch, deep fry in heated oil to crispy, drain and stir with mixed seasonings ❷ , pour over lettuce. Serve.

Tips

· The prawns added can increase the quantity of lobster meat.
· This dish, which we call "pineapple shrimp balls" was inspired by a Hong Kong chef's creation of "mayonnaise sesame shrimp," the term of which plays on a way Cantonese people refer to mayonnaise.

龍蝦粉絲煲

Lobster in Casserole

<table>

使用材料	活龍蝦	1 隻（約 1 斤多重）	鹽	¼ 茶匙
	葱	3 支	糖	½ 茶匙
	薑	10 片	黑胡椒粉	¼ 茶匙
	粉絲	2 把	水	2 杯
	青蒜絲	少許		

</table>

調味料

酒	1 湯匙
醬油	2 湯匙

01 龍蝦處理後，分斬成 10 ～ 12 小塊，沾上少許麵粉，用熱油煎過。葱 3 支切段。

02 另燒熱 3 湯匙油，爆香葱段及薑片，放下龍蝦及調味料，大火煮滾。

03 粉絲用冷水泡軟，剪短。起油鍋爆香葱段，淋醬油 1 湯匙、水 1 杯，放入粉絲煮至半透明，倒入砂鍋中，再倒下②龍蝦料，大火煮 1 分鐘，撒下青蒜絲即可。

INGREDIENTS •

1 Live lobster, 3 stalks Green onion, sections, 10 Ginger slices, 2 bundles Mung bean thread, some Green garlic shreds.

SEASONINGS

1tbsp. Wine, 2tbsp. Soy sauce, ¼tsp. Salt, ½tsp. Sugar, ¼tsp. Pepper, 2cups Water.

STEP BY STEP •

01 Kill the live lobster, cut into 10 pieces. Coat with flour & fry in hot oil until browned, remove.

02 Heat 3tbsp. of oil to stir fry ginger & green onion, add lobster, then add in seasonings, bring to a boil.

03 Soak mung bean threads to soft, cut into sections about 4" long, put into a casserole, pour the lobster & soup in, cook over medium heat until the mung bean threads turns transparent and the soup reduced, turn off the heat, sprinkle the green garlic shreds over. Serve.

炸蛋黃蝦排

Deep Fried Prawns with Egg Batter

使用材料	草蝦或小明蝦	10 隻	葱	1 支
	火腿屑	2 湯匙	薑	1 片
	香菜葉	10 片		
			蛋黃糊	
	醃蝦料		蛋黃	2 個
	酒	1 茶匙	鹽	少許
	鹽	¼ 茶匙	麵粉	2 湯匙
	糖、麻油	各少許		
	太白粉	1 茶匙		

01 蝦去殼僅留尾部,抽除砂筋後,用醃蝦料醃 10 分鐘、調好蛋黃糊備用。

02 將醃過之蝦肉撒少許太白粉後,每隻沾上蛋黃糊並在兩面撒下火腿屑和香菜葉片。

03 將蝦投入熱油中炸熟,瀝淨油後,由背部平刀片切成兩半,排列在盤中,附上花椒鹽及番茄醬上桌。

INGREDIENTS

10 pieces Prawn, 2tbsp. Ham (chopped), 10 pieces Chinese parsley leaf.

SEASONINGS

❶ 1tsp. Wine, ¼tsp. Salt, ½tsp. Sugar, ½tsp. Sesame oil, 1tsp. Cornstarch, 1 stalk Green onion, 1 slice Ginger .

❷ 2 Egg yolk, ½tsp. Salt, 2tbsp. Flour.

STEP BY STEP

01 Shell the prawns, (leave the tail shell on), marinate with ❶ for 10 minutes. Mix ❷ evenly.

02 Pat prawns with cornstarch, coat with ❷ batter, then sprinkle the ham and Chinese parsley on both sides.

03 Deep fry prawns in hot oil to golden brown, drain and slice horizontally into halves, arrange on a plate. Serve with ketchup and mixed brown pepper corn powder with salt.

四色鳳尾蝦

Phenix Tail Prawns

01 蝦剝殼，僅留尾部，由蝦背剖開成一大片，並用刀面拍平，用醃蝦料拌醃 10 分鐘。

02 香菇泡軟，連汁與蒸香菇料蒸 15 分鐘，取出切絲。蛋做成蛋皮切絲。豌豆莢切絲。豆腐半盒先對半切為二後，再橫片 2 刀，共 6 片，平鋪盤底，上面撒上少許鹽。

03 4 種絲料各取 3、4 條，橫放在蝦身上，再將蝦尾彎曲一下，排在豆腐上，大火蒸 4 分鐘，取出後湯汁泌到鍋中，再加清湯半杯及鹽少許，煮滾勾芡，澆到蝦上即可。

Tips

蝦腹部之白筋不可抽斷，否則蝦尾熟後就不會彎曲了。

INGREDIENTS •

6 pieces Prawn, 2 Black mushrooms, 2tbsp. Ham shreds, 1 Egg, some Pea pods, ½pack Bean curd, ½ cup soup stock, some salt.

SEASONINGS

❶ ¼tsp. Salt, 1tsp. Cornstarch, 1tsp. Wine.

❷ 1tbsp. Soy sauce, ¼tsp. Sugar, 1tsp. Oil.

STEP BY STEP •

01 Shell the prawns, (leave the tail shell on), cut the back of prawns to let prawn spread to a piece, pat with the cleaver, marinate with ❶ for 10 minutes.

02 Soak the mushroom in a bowl with water to soft, add ❷ and steam for 15 minutes, drain and cut into fine shreds. Beat the egg and fry into a sheet, cut into shreds. Shred the pea pods. Cut bean curd into 6 thick slices, arrange on a plate, sprinkle some salt on.

03 Put 4 strings of mushroom, ham, pea pods & egg on belly part of prawn, bending the tail to the head side, to make a curly shape. Place one prawn on each bean curd slice. Steam for 4 minutes. Thicken the steamed soup and soup stock, add some salt, mix and pour over prawns.

Tips

Do not remove the white vein from the belly part or the tail will not become curly after steamed.

香酥小蝦捲

Crispy Shrimp Rolls

01　小蝦仁用少許鹽抓洗過後，沖水並擦乾，每隻切為二段，加入醃蝦料拌醃一下；荸薺切小粒；肥肉煮熟，冷後切小粒；蔥切成屑。

02　將①項材料置碗中，加入魚漿及調味料拌勻，豆腐衣切成 3 寸寬長方形，每張包入蝦料，捲成如中指般長條（豆腐衣上塗少許麵粉糊），收口處用麵粉糊黏住。

03　蝦捲用熱油、小火炸熟至外皮酥脆為止，撈出裝盤，可附花椒鹽或番茄醬沾食。

INGREDIENTS •

150g. Shrimp (shelled), 80g. Fish paste, 4 pieces Water chestnut, some Pork fat, 1 stalk Green onion, chopped, 2 pieces Dried bean curd sheet, some flour paste.

SEASONINGS

❶ ¼tsp. Salt, 1tbsp Egg white, 1tsp. Cornstarch.

❷ ½tsp. Salt, ¼tsp. Sugar, ½tsp. Sesame oil, Five-spicy powder.

STEP BY STEP •

01　Rinse and clean the shrimp, drain and cut each into two pieces, marinate with ❶ for a while. Cut water chestnuts into small pieces. Boil the pork fat and cut into small cubes. Chop green onion.

02　Mix shrimp, water chestnuts, pork fat, chopped green onion, fish paste, and ❷ , mix well. Cut dried bean curd sheet into 3" wide rectangles to wrap in the shrimp mixture, roll into a thumb-size cylinder. Seal edges with flour paste.

03　Deep fry shrimp rolls over medium heat to golden brown. Serve with ketchup or brown pepper corn powder with salt.

醬燒蝦

Stewed Shrimps with Brown Sauce

使用材料				
	大草蝦或蘆蝦	10 隻	水	3 湯匙
	太白粉	1 湯匙	糖、醋	各 1 茶匙
	大蒜片	15 片	麻油	½ 茶匙
	陳皮屑	1 茶匙	胡椒粉	少許
	紅辣椒屑、柳橙皮屑	1 茶匙		
	綜合調味料			
	蠔油、醬油	各 ½ 湯匙		
	酒	1 湯匙		

步驟說明

01　每隻蝦剪除鬚與腳後，由腹部剖開一刀，並用刀面拍平一點，撒下太白粉拌勻，用熱油炸熟撈起。

02　用1湯匙油爆香大蒜片，淋綜合調味料煮滾，放入陳皮屑與蝦，燒煮至汁乾，裝盤後撒下紅辣椒屑及柳橙屑即可。

醬燒與紅燒不同，醬燒的菜色必須將醬汁燒煮到汁收乾，而紅燒的菜仍帶有湯汁。

INGREDIENTS •

10 pieces prawns, 1tbsp. Cornstarch, 15 pieces Garlic slice,1tsp. Dried tangerine peel, 1tsp. Red chili, chopped, 1tsp. Fresh orange peel, chopped.

SEASONINGS

½tbsp. Oyster sauce, ½tbsp. Soy sauce, 1tbsp. Wine, 3tbsp. Water, 1tsp. Sugar, 1tsp. Vinegar, ½tsp. Sesame oil, a pinch of Pepper.

STEP BY STEP •

01　Trim the shrimp, score a cut on the belly part, pat with the side of a cleaver. Coat with cornstarch and deep fry in hot oil to done, drain.

02　Heat 1tbsp. of oil to stir fry garlic slices, add seasonings, bring to a boil, add tangerine peel & shrimp, cook until sauce reduced. Sprinkle the chopped fresh orange peel & red chili. Serve.

Tips

"Dry sauce" is different from sauce in foods that are braised. For dishes cooked with "dry sauce," the item must be cooked until the sauce dries, whereas dishes that are braised still come with broth.

芥辣鮮蝦絲

Shrimp Shreds with Mustard Sauce

62

01 蝦去殼後洗淨，每隻橫切成 2 片，並劃一刀口在蝦肉上，全部用蛋白半個和鹽少許拌勻，再在太白粉中沾裹，並用力壓成一大片。

02 將綜合調味料調勻；菠菜切短；木耳切寬絲，一起在開水中川燙一下，瀝乾後放盤中，倒入一半已調勻之綜合調味料拌合。

03 蝦片也投入開水中川燙至熟，瀝乾後切成絲狀，用剩餘之綜合調味料拌勻，堆放在菜料上即可。

INGREDIENTS •

6 Prawns, 150g. Green vegetable, ½ cup fungus, 1cup Cornstarch, 1tbsp. Egg white.

SEASONINGS

1tbsp. chopped Green onion, 1tbsp. Chinese parsey, 1tbsp. Mustard paste, 1tbsp. Sesame seed paste, ½tbsp. Soy sauce, ½tbsp. Sesame oil, ¼tsp. Salt, ½tsp. Sugar, a pinch of Pepper.

STEP BY STEP •

01 Shell the prawns, slice each into halves, score one cut on each half, mix with egg white & salt, then coat with cornstarch, press into a large flat piece.

02 Mix the seasonings. Blanch vegetable and fungus, drain and mix with half amount of seasonings, Put on a plate.

03 Blanch prawns in boiling water, drain and cut into shreds, mix with another half of seasonings, put on top of the vegetables. Serve.

雞絲燴翅

Stewed Shark's Fin with Chicken Shreds

使用材料					
水發魚翅	4 兩		太白粉	1 茶匙
雞胸肉	3 兩		水	2 茶匙
綠豆芽	1 小碟		鹽	¼ 茶匙
葱	2 支			
薑	3 片		綜合調味料	
高湯	6 杯		酒、醬油	各 1 湯匙
香菜	少許		鹽	½ 茶匙
			太白粉水	1 ½ 湯匙
醃雞料				
蛋白	1 湯匙			

01　魚翅用冷水及蔥、薑、酒煮 10 分鐘。去除腥味，另換高湯 1 ½ 杯煨煮至軟。

02　雞肉順紋切細絲，用醃雞料醃 15 分鐘，過油泡炒至熟。

03　豆芽摘去頭尾，用少量油炒熟；香菜洗淨、切短。

04　用 1 湯匙油煎香蔥、薑，淋入調味料中之酒及高湯，煮 3 分鐘，挑出蔥、薑，將魚翅下鍋，用醬油及鹽調味，勾芡後將雞絲拌入即熄火，分裝小碗或盅內，附銀芽及香菜上桌。

INGREDIENTS

150g. Shark's fin, 120g. Chicken breast, 1cup Mung bean sprouts, 2 stalks Green onion, 3 pieces Ginger slice, 6 cups Soup stock, a little of Chinese Parsley.

SEASONINGS

❶ 1tbsp. Egg white, 1tsp. Cornstarch, 2tsp. Water, ¼tsp. Salt.
❷ 1tbsp. Wine, 1tbsp. Soy sauce, ½tsp. Salt, 1 ½tbsp. Cornstarch paste.

STEP BY STEP

01　Cook shark's fin in cold water with green onion, ginger and wine for 10 minutes, drain and cook with 1 ½cups of soup stock until soft.

02　Shred the chicken, marinate with sauce ❶. Fry in warm oil till done, drain.

03　Trim off 2 ends from bean sprouts, stir fry with heated oil. Rinse Chinese parsley, cut into small sections.

04　Heat 1tbsp. of oil to fry green onion and ginger, add wine & soup stock, cook for 3 minutes, discard the green onion & ginger. Add shark's fin and seasonings ❷, mix in the chicken shreds. Pour into a soup bowl, serve with bean sprouts & Chinese parsley.

魚翅石榴包

Sharks Fin in Pomegranate Package

步
驟
說
明

01 魚翅放鍋中加水、葱、薑及酒，小火煮 10 分鐘，去腥味後撈出，另用 1 杯高湯煨煮至軟。雞胸肉切細絲，用少許太白粉加水及鹽拌醃一下；香菇泡軟、切絲；金菇去除尾端切短；芹菜燙軟。

02 熱油 2 湯匙爆香葱、薑，並將雞絲炒散，再放①項中各種絲料，並加調味料及魚翅炒勻。

03 蛋白打散，加入其他蛋白皮料（水和太白粉要先調勻），打勻後再用小網過濾一次。鍋中刷少許油，將蛋白分別煎成約 6 片、4 寸大小的圓薄皮。

04 用蛋白皮來包魚翅料，收口用燙軟之細芹菜紮好，分別裝入小碗中，入鍋蒸 4 分鐘即可，每碗中加煮滾且調味之高湯即可供食。

INGREDIENTS •┄┄┄┄┄┄

120g. Shark's fin, ½ Chicken breast, 4 pieces Black mushroom, a few stalk Celery, 50g. Needle mushroom, Ginger, Green onion, chopped, (each a little), a little of Cornstarch paste, 4 cups Soup stock.

EGG SKIN

4 Eggs white, 1tsp. Cornstarch, 1tbsp. Water, ¼tsp. Salt.

SEASONINGS

1tsp. Wine, ¼tsp. Salt, a pinch of Sugar, a pinch of Pepper, some Cornstarch paste.

STEP BY STEP •┄┄┄┄┄┄

01 Cook shank's fin in water with ginger, green onion and wine for 10 minutes, drain. Simmer with 1 cup of chicken soup stock until soft. Shred the chicken breast, marinate with cornstarch paste and salt. Shred the soaked black mushroom. Cut off the end of needle mushroom. Blanch the celery.

02 Heat 2tbsp. of oil, stir fry green onion, ginger, chicken and all vegetables, add seasonings and mix well with shark's fin.

03 Beat eggs white and mix all egg skin materials(mix cornstarch with water first, then add to egg white.) Brush a little of oil on wok, fry 6 round thin egg sheets (about 4" diameter round).

04 Use egg sheet to wrap shark's fin mixture, fasten with blanched celery strings. Put each in a bowl. Steam for 4 minutes. Add in boiling soup stock (seasoned). Serve.

鮑魚火腿扣通粉

Mold Abalone and Ham with Macaroni

使用材料

罐頭鮑魚	半罐
中國火腿絲	半杯
洋葱丁	2 湯匙
通心粉	1 ½ 杯
麵粉	2 湯匙
高湯	3 杯

調味料

鹽	½ 茶匙
胡椒粉	少許
牛奶	3 湯匙

01　鮑魚、火腿均切絲；通心粉放滾水中，以中火煮熟（如用中國火腿需蒸約 20 分鐘後切絲）。

02　取一只蒸碗，底層排入鮑魚及火腿絲，中間填入通心粉，撒 ¼ 茶匙鹽及半杯水，蒸 15 分鐘，扣在大盤中。

03　用 3 湯匙油將洋蔥丁炒軟，加入麵粉同炒，慢慢淋入高湯攪勻，煮滾後撈棄洋蔥，加鹽調味，熄火後加入牛奶及胡椒粉，拌勻後淋到鮑魚通心粉上。

Tips

可將鮑魚、火腿、通心粉與麵糊拌勻後裝入烤碗中，撒少許起司粉烤熟，但是烤的麵粉糊要做得濃稠些。

INGREDIENTS •

½tsp. Can Abalone, ½ cup Chinese ham shreds, 2tbsp. Onion, chopped, 1 ½cups Macaroni, 2tbsp. Flour, 3cups Soup stock.

SEASONINGS

½t Salt, a pinch of Pepper, 3tbsp. Milk.

STEP BY STEP •

01　Shred the abalone. Steam the Chinese ham for 20 minutes, cut into shreds. Boil the macaroni.

02　Put abalone and ham strings on bottom of a bowl, stuff center with macaroni, add ¼tsp. salt & ½cup of water. Steam for 15 minutes, reverse the bowl and let stand on a plate.

03　Heat 3tbsp. of oil to stir fry onion and flour, add in soup stock and mix evenly, discard onion, season with salt, pepper and milk. Mix and pour over abalone. Serve.

Tips

You may mix the ingredients with cream sauce and bake it.

鮑魚燴三白

Braised Abalone with Three Kinds of Ingredients

使用材料

鮑魚	半罐
白蘆筍	12 支
雞胸肉	半個
竹笙	10 條
高湯	4 杯

調味料

鹽	酌量
太白粉水	2 茶匙
雞油	少許

01　將雞胸肉切成薄片，用 1 茶匙太白粉水及鹽少許拌勻，醃十餘分鐘，用 1 杯油將雞肉過油泡熟後瀝出。

02　蘆筍對切為二，蒸 8 分鐘排入盤中墊底。

03　竹笙泡軟，多沖洗幾次，最後用開水燙一下，切成 5 公分長段，再用 1 杯高湯蒸 10 分鐘。鮑魚切薄片留用。

04　將 3 杯高湯煮滾後加鹽調味，放下雞片、竹笙，一滾即勾芡，盛入大盤中，鍋中留半量湯汁，放入鮑魚片，再一滾即全部倒入大盤內，淋少許雞油便成。

Tips

罐頭中之鮑魚湯汁可包括在高湯中使用之。

INGREDIENTS •

½ can Abalone, 12 stalks white asparagus, ½ Chicken breast, 10 pieces Bamboo mushroom, 4 cups Soup stock.

SEASONINGS

a pinch of Salt, 2tsp. Cornstarch paste, some Chicken grease.

STEP BY STEP •

01　Slice chicken breast, marinate with cornstarch paste & salt for 10 minutes, then stir fry in 1 cup of oil(160°C) to done, drain immediately.

02　Cut asparagus into 2 sections, steam for 8 minutes and put on a plate.

03　Soak the dried bamboo mushrooms, change water for several times while soaking. Blanch in boiling water, cut into 5 cm sections. Add 1cup of soup stock and steam for 10 minutes. Slice abalone thinly.

04　Cook 3 cups of soup stock, add salt, chicken, bamboo mushrooms, bring to a boil, thicken with cornstarch paste, remove to the plate. Keep half amount of soup stock in wok, add abalone, bring to a boil again, Pour over the chicken. Drizzle with chicken grease. Serve.

Tips

The canned abalone soup can be included in 3cups of soup stock.

銀紙蒸鮮貝

Steamed Scallops

使用材料		
新鮮干貝	6 粒	
葱、薑絲、胡蘿蔔絲	酌量	
鋁箔紙盒	2 張	
芥蘭菜葉	3 片	

調味料

水	¾ 杯
酒	½ 湯匙

鹽	¼ 茶匙
油	½ 湯匙
胡椒粉	少許

01　干貝橫剖爲二，鋁箔紙盒內（可用鋁箔紙折疊）鋪上燙過的芥蘭菜葉，每個中間放入干貝 6 片，再在上面撒下蔥、薑絲及胡蘿蔔絲各少許。

02　調味料調妥後注入干貝中，再放入蒸籠中蒸 3 分鐘即成，也可放在烤箱中烤熟。

可用蛤蜊、明蝦、螃蟹、蚵等其他海鮮類來代替新鮮干貝。

INGREDIENTS ·

6 pieces Fresh scallop, a little amount of shredded Green onion, Ginger and Carrot, 2 pieces Aluminum foil, 3 pieces mustard green leaf.

SEASONINGS

¾ cup Water, ½tbsp. Wine, ¼tsp. Salt, ½tbsp. Oil, a pinch of Pepper.

STEP BY STEP ·

01　Slice scallop into 2 slices. Place the blanched mustard green leaves on inside of the aluminum foil box. Put scallop slices in and cover with ginger and green onion, carrot each a little bit.

02　Mix the seasonings and pour into the scallop boxes. Arrange on a steamer and steam over high heat for 3 minutes. Serve. (You may also bake in an oven.)

Tips

The crabs, clams, prawns, oyster or other kinds of seafood can be also cooked with this way.

碧綠玉筍帶子

Fresh Scallops with Baby Corn & Broccoli

01　將干貝片切成 0.3 公分厚的片狀，全部用醃料拌醃半小時。香菇泡軟切成小片。

02　將青花菜分割成小朵，和對切後的玉米筍全部用滾水燙熟（水中需加油 1 湯匙及鹽半茶匙），撈出排在盤中。

03　將水燒至四分熱，放下干貝片，小火泡至熟，撈出。

04　用 2 湯匙油，起油鍋爆香蔥、薑及香菇片，放入熟胡蘿蔔片及干貝，淋下綜合調味料，大火拌炒均勻即裝入盤中。

INGREDIENTS •

6 pieces Fresh scallop, 1 stalk Green onion, 2 Ginger slices, 4 Black mushrooms, 1 Broccoli, 10 Baby corns, 15 pieces Carrot slice, cooked.

SEASONINGS

❶ ⅓ tsp. Salt, ½ tbsp. Wine, a pinch of Pepper, 1tbsp. Cornstarch.
❷ ½ tbsp. Wine, ½ tbsp. Light colored soy sauce, ¼ tsp. Salt, a pinch of Sugar & Pepper, ¼ tbsp. Sesame oil.

STEP BY STEP •

01　Cut fresh scallop into 0.3 cm thick slices, marinate with ❶ for ½ hour. Soak the black mushroom to soft then cut each into small pieces.

02　Trim the broccoli. Cut the baby corn into two pieces. Blanch these two in boiling water (add 1tbsp. of oil & ½tsp. of salt in water), drain and arrange on the plate.

03　Blanch scallop in warm water(about 80°C) over low heat for ½ minute, drain.

04　Heat 2tbsp. of oil to fry green onion, ginger and black mushroom, then add carrots & scallops, pour ❷ in, stir well over high heat and serve.

鍋貼干貝酥

Fried Scallop Cake

使用材料

雞胸肉或雞柳	3兩	鹽	¼ 茶匙
干貝	3粒	太白粉	1 茶匙
芥蘭菜葉	酌量	水	1 茶匙

拌雞料

葱薑水	2湯匙
蛋白	2個

蛋汁料

蛋	2個

01 雞肉用刀刮成泥狀，加入蔥薑水調開，將蛋白慢慢加入，調成糊狀，便是雞茸糊。干貝蒸約 20 分鐘至軟，涼後撕成絲。

02 蛋汁料打散調勻（太白粉和水須先調勻再加入蛋汁中）。鍋中塗油將蛋汁倒下，攤煎成 15 公分左右的蛋皮，在蛋汁半凝固時，即將雞茸糊均勻淋在蛋皮上，再將干貝絲撒下。

03 沿鍋邊淋入油半湯匙，小火慢煎至凝固狀態時，小心翻面再略煎一下即可，切塊排盤。芥蘭菜葉可切絲墊在盤底。

INGREDIENTS •

120g. Chicken breast, 3 pieces dried Scallop, a little of Mustard green leaves.

SEASONINGS

❶ 2tbsp. Green onion & Ginger juice, 2 Eggs white.
❷ 2 Eggs, ¼tsp. Salt, 1tsp. Cornstarch, 1tsp. Water.

STEP BY STEP •

01 Smash the chicken, mix well with ❶ . Steam the dried scallops for 20 minutes with water and tear into strings after cools.

02 Mix ❷ . Heat a little of oil to fry the egg mixture, spread into a 15cm diameter round cake, spread chicken mixture on while the egg mixture is half cooked and then arrange scallop on top.

03 Add ½tbsp. of oil around the edge of the wok. Fry over low heat until chicken mixture solidified. Turnover and fry to done. Cut scallop cake into pieces, arrange on a plate. You may shred some mustard green leaves on the bottom of the plate.

金盤雙魷

Two kinds of Squid in Golden Plate

<table>
<tbody>
<tr><td rowspan="13">使用材料</td></tr>
</tbody>
</table>

使用材料

餛飩皮	20 張	糖	½ 湯匙
鮮魷及水發魷魚	各 1 條	酒	1 湯匙
花椒粒	½ 湯匙	鹽	¼ 茶匙
乾辣椒	10 支	醋	2 茶匙
麵糊	適量	水	1 湯匙
		太白粉	1 茶匙
綜合調味料		麻油	1 茶匙
醬油	1 湯匙		

01　分別在鮮魷及水發魷魚內部切交叉切口,再分割成 1 寸多大小。用滾水川燙過,瀝出。

02　將 20 張餛飩皮鋪排在漏勺上,每張皮交接處需塗少許麵糊,使其黏住。用另一漏勺沾油後壓住,全部放入熱油中炸熟,小心取下餛飩皮金盤,放在盤中(兩把漏勺均需先在油中沾一下,以免黏住取不下來)。

03　用 2 湯匙油將花椒粒煎香後撈棄,放入乾辣椒段(切 1 寸長)炸香,放下雙魷花及綜合調味料,大火拌炒均勻,盛到金盤中。

INGREDIENTS ·

20 pieces Won-Ton wrapper, 1 Squid (Dried & Soaked), 1 Fresh squid, ½tbsp. Brown pepper corn, 10 pieces Dried red chili, a little of flour paste.

SEASONINGS

1tbsp. Soy sauce, ½tbsp. Sugar, 1tbsp. Wine, ¼tsp. Salt, 2tsp. Vinegar, 1tsp. water, 1tsp. Cornstarch , 1tsp. Sesame oil.

STEP BY STEP ·

01　Slice crisscross on inside of fresh squid and soaked dried squid, then cut each into 1" squares, blanch in boiling water. Drain.

02　Brush oil on a round frying strainer. Place the Won-Ton wrapper one by one on the strainer, stick on each other with flour paste, then press down the strainer with another strainer (dip in some oil too), deep fry in hot oil for 1 minute, remove the golden plate carefully, put on a plate.

03　Heat 2tbsp. of oil to fry brown pepper corn for a while, discard. Use this oil to fry dry red chili, (cut into 1" long), then add 2 squids and mixed seasonings in, stir over high heat, mix evenly and pour into the golden plate. Serve hot.

白灼鮮魷

Blanched Fresh Squid

使用材料

新鮮魷魚	2 條
葱	1 支
薑	2 片
酒	1 湯匙

薑醋汁

薑末	½ 湯匙
白醋、醬油、麻油	各 1 湯匙

芥末汁

芝麻醬、醬油	各 1 湯匙
糖	½ 茶匙
芥末粉、麻油	各 1 茶匙

麻辣汁

葱、蒜屑	各 ½ 湯匙
辣豆瓣醬、醬油	各 ½ 湯匙
糖、麻油	各 1 茶匙

步驟說明

01 將魷魚外皮剝除洗淨，先橫切成 1 ½ 寸寬段，每段再取 3 寸長，使每一塊均為 1 ½X3 寸大小，每一塊前一半切交叉刀紋，後一半切直條刀口。

02 燒半鍋水，加蔥、薑及酒，滾過 2 分鐘後將切花之魷魚放下，川燙 10 秒鐘即撈起，瀝乾排入盤中。

03 備 3 只小碗，分別或選擇自己喜愛的口味調好佐料，和鮮魷一起上桌。

INGREDIENTS

2 Fresh squids, 1 stalk Green onion, 2 Ginger slices, 1tbsp. Wine.

GINGER VINEGAR SAUCE

½tbsp. Smashed ginger, 1tbsp. White vinegar, 1tbsp. Soy sauce, 1tbsp. Sesame oil.

MUSTARD SAUCE

1tbsp. Sesame seed paste, 1tbsp. Soy sauce, ½tsp. Sugar, 1tsp. Mustard powder, 1tsp. Sesame oil.

HOT SPICY SAUCE

½tbsp. Chopped green onion, ½tbsp. Chopped garlic, ½tbsp. Hot bean paste, ½tbsp. Soy sauce, 1tsp. Sugar, 1tsp. Sesame oil.

STEP BY STEP

01 Trim the squid, cut (along the grain) into 1 ½" wide sections, then cut each section into 1 ½"x3" rectangles. Cut half part of each rectangle crisscross, then cut the other half part into strings.

02 Add ginger, green onion with wine in boiling water, boil for 2 minutes. Blanch squid for 10 seconds till cured, drain and arrange on a plate.

03 Serve with three kinds of sauce or any one you prefer.

果粒溜雙鮮

Squid and Shrimp with Fruits

使用材料		
新鮮墨魚（即花枝）	1 隻	
小型草蝦	8 隻	
鳳梨丁	少許	
青椒絲、紅辣椒絲	各少許	
美乃滋	½ 杯	
牛奶	1 湯匙	
白芝麻	½ 湯匙	
玉米粉	半杯	

醃料	
鹽	¼ 茶匙
酒	½ 茶匙
胡椒粉	¼ 茶匙
蛋白	1 湯匙
太白粉	½ 茶匙

01 墨魚剝去外皮後切成 1 寸長的粗條。草蝦剝殼，洗淨擦乾，背部劃一刀口。兩者一起用醃料醃 10 分鐘，沾上玉米粉。

02 美乃滋放大碗中，加入牛奶調勻，放下鳳梨丁、青、紅椒絲備用。

03 燒熱炸油，放入墨魚及草蝦，大火炸酥，瀝出後裝入②項的大碗中，輕輕翻動拌勻，裝入盤中，撒下炒香的白芝麻，即可享用。

INGREDIENTS •

1 Cuttlefish, 8 Prawns, Pineapple, diced, Green pepper and red chili shreds, a little of each, ½ cup Mayonnaise, 1tbsp. Milk, ½tbsp. Sesame seeds, ½ cup Cornstarch.

SEASONINGS

¼tsp. Salt, ½tsp. Wine, ¼tsp. Pepper, 1tbsp. Egg white, ½tsp. Cornstarch.

STEP BY STEP •

01 Clean and cut cuttlefish into 1" long strips. Shell the prawns, marinate with seasonings for 10 minutes, coat with cornstarch.

02 Mix mayonnaise and milk, add pineapple and green pepper & red chili shreds in.

03 Deep-fry cuttlefish and prawns over high heat to golden brown. Drain and mix with②. Remove to a plate and sprinkle the sesame seeds over. Serve.

咖哩鮮蟹煲

Curry Sauce Crab in Casserole

使用材料				
	青蟹或其他鮮蟹	2 隻	酒	1 湯匙
	粉絲	2 把	鹽	¼ 茶匙
	大蒜屑	1 湯匙	糖	¼ 茶匙
	洋蔥丁	半杯	高湯或水	2 杯
	青蒜絲	少許		
	調味料			
	咖哩粉	1 ½ 湯匙		

01 將蟹切成小塊，用 4 湯匙熱油炒熟。

02 另用 2 湯匙油炒香大蒜屑及洋蔥丁，並加入咖哩粉小火炒香，淋下酒，加入蟹塊拌炒片刻，調味後注入高湯，大火燒煮至滾。

03 砂鍋內放下泡軟之粉絲，倒入②項之湯汁及蟹（水若太少可酌量增加）。蓋妥蓋子，小火繼續燒 3 分鐘。撒下青蒜絲或蔥花，淋少許油即可上桌分食。

INGREDIENTS •

2 Crabs, 2 bundles Mung bean threads, 1tbsp. Garlic, chopped, ½ cup Onion, chopped, a little of Green garlic shreds.

SEASONINGS

1 ½tbsp. Curry powder, 1tbsp. Wine, ¼tsp. Salt, ¼tsp. Sugar, 2cups Soup stock or water.

STEP BY STEP •

01 Cut crab into small pieces, stir fry with 4tbsp. of heated oil.

02 Heat 2tbsp. of oil to stir fry garlic & onion, add in curry powder, stir fried over low heat until fragrant, add wine and crab in, stir fry for a while, finally pour the soup stock in, bring to a boil.

03 Put soaked mung bean threads in a casserole, pour the crab and sauce in the casserole, cover and simmer for 3 minutes, sprinkle green garlic or green onion. Serve.

椒鹽焗花蟹

Salt & Pepper Crab

使用材料		
花蟹或海蟹	2 隻
太白粉	2 湯匙
檸檬片	3 片
茶水	1 杯

調味料

鹽	¼ 茶匙
五香粉、胡椒粉	各少許

01 將蟹蓋打開，除去腮等，分割成小塊，用太白粉拌勻，投入熱油中炸熟。

02 乾淨鍋子燒熱後，放下炸蟹塊，緩緩撒下已混合之調味料，不停拌炒至均勻為止。

03 裝盤後與泡了檸檬片的茶水（食後洗手用）同時上桌。

 Tips

也可將炸過之蟹塊用蔥、薑、蒜末烹過，口味較重。

INGREDIENTS •

2 Crabs, 2tbsp. Cornstarch, 3 pieces Lemon slice, 1cup of Tea.

SEASONINGS

¼tsp. Salt, a pinch of Five-spicy powder, a pinch of Pepper.

STEP BY STEP •

01 Open the crab lid, remove dirt, cut into 4 pieces, mix with cornstarch. Deep fry in heated oil till done.

02 Heat the wok, add crabs and seasonings, stir evenly.

03 Serve with Tea (with lemon added) for hand washing.

 Tips

You may stir fry the frying crabs with chopped garlic, green onion & ginger, the taste will be stronger.

麻辣海參

Spicy Sea Cucumber

01 海參先放鍋中，加蔥 2 支、薑 2 片、酒 1 湯匙，及冷水 3 杯，煮至適當軟度，取出。先每條直切成兩半，再切成半寸多厚斜片。

02 熟豬肉切小厚片；青蒜切斜段；黃瓜剖成四半，切斜段備用。

03 用 1 湯匙油炒透肉片，加入辣豆瓣醬同炒，並加水、醬油、糖、酒等調味料，放入海參及黃瓜片，燒煮 3 ～ 4 分鐘。

04 加入青蒜段拌合後勾芡，淋下麻油、花椒粉即可裝盤。

INGREDIENTS •

2 pieces Sea cucumber, 120g. Cooked pork, 1 Cucumber, 1 stalk Green garlic.

SEASONINGS

1tbsp. Hot bean paste, 1cup Soup stock, ½tbsp. Soy sauce, 1tsp. Sugar, 1tsp. Wine, 1tbsp. Cornstarch paste, 1tsp. Sesame oil, ¼tsp. Brown pepper corn powder.

STEP BY STEP •

01 Put sea cucumber in a wok, add 3 cups of cold water, 2 stalk of green onion, 2 slices ginger & 1tbsp. wine, cook over medium heat to moderate soft, drain. Cut each to two pieces, then diagonally into 4 slices.

02 Slice the cooked pork, shred the green garlic, cut cucumber into slices.

03 Heat 1tbsp. of oil, stir fry pork, add hot bean paste and all seasonings, finally add sea cucumber and cucumber slices, cook for 3~4 minutes.

04 Mix with green garlic shreds, thicken with cornstarch paste, splash sesame oil, sprinkle brown pepper corn powder. Serve.

如意海參

Stuffed Sea Cucumber with Chicken

海參	3 條		鹽	¼ 茶匙
雞胸肉	3 兩		太白粉	2 茶匙
洋火腿	3 條			
玉米筍、青花菜	各適量		**調味料**	
玻璃紙	3 張		高湯	1 ½ 杯
			醬油、酒	各 ½ 湯匙
雞茸料			鹽、胡椒粉	各少許
酒	1 茶匙		太白粉水	2 茶匙
葱薑水	1 茶匙		麻油	½ 茶匙

01　雞胸肉用刀剁成雞茸，加入雞茸料拌勻。

02　海參洗淨腸砂，放入鍋中出水（可參考麻辣海參 p.89）至軟。擦乾內部，撒入乾太白粉，釀入雞茸，雞茸中央放 1 條如筷子般粗細的火腿條（火腿要先沾上太白粉），用玻璃紙包捲成長條型，蒸 10 分鐘左右。

03　取出海參包，待冷卻後，拆除玻璃紙，切成薄圓片，排入盤中。

04　用 1 湯匙油煎香葱、薑，注入高湯並調味，勾芡後滴入麻油，再淋到海參片上。用燙過之玉米筍及青花菜圍飾即可。

Tips

海參夠大時，可完全包裹住雞茸，切片後即成金錢狀，名爲「金錢海參」。

INGREDIENTS •

3 pieces Sea cucumber, 120g. Chicken breast, 3 pieces Ham (cut into chopstick sized stick), 1cup Baby corn, 1cup Broccoli, 3 pieces Cellophane paper.

SEASONINGS

❶ 1tsp. Wine, 1tsp. Green onion & Ginger juice, ¼tsp. Salt, 2tsp. Cornstarch.

❷ 1 ½cups Soup stock, ½tbsp. Wine, ½tbsp. Soy sauce, a pinch of Salt, a pinch of Pepper, 2tsp. Cornstarch paste, ½tsp. Sesame oil.

STEP BY STEP •

01　Chop and smash the chicken, mix well with ❶.

02　Rinse and clean the sea cucumber, discard the intestine, cook with water, ginger, green onion and wine over medium heat for 3~5 minutes, drain. Pat dry with a towel, sprinkle cornstarch, stuff with the chicken meat, insert a stick of ham, wrap and roll tightly with cellophane paper. Steam for 10 minutes.

03　Take out the sea cucumber, discard cellophane paper, cut into 1" slices, arrange on a plate.

04　Heat 1tbsp. of oil to stir fry green onion & ginger, add soup stock, soy sauce and wine, season with salt, and pepper, thicken with cornstarch paste, splash sesame oil, pour over sea cucumber, decorate with baby corn & broccoli.

If the Sea cucumber is big enough, it can coat the chicken meat all over, then it will become a coin shape after slice it.

香汁蝴蝶魚

Butterfly Fish

使用材料

小草魚	1 條
蔥、薑絲	各 1 湯匙
玉米粉	½ 杯
麵粉	少許

醃魚料

鹽	⅓ 茶匙
酒	½ 湯匙
胡椒粉	少許
蛋白	半個
太白粉	1 茶匙

綜合調味料

糖	2 湯匙
醋	1 湯匙
水	2 湯匙
淡色醬油	1 湯匙
鹽	¼ 茶匙
桔子汁	3 湯匙
太白粉	½ 湯匙
麻油	1 茶匙

01　魚頭及尾部切下，用少許鹽、酒抹一下，裹上麵粉。

02　魚肉去骨後，切成蝴蝶薄片（連皮第一刀不切斷，第二刀才切斷），用醃魚料醃 10 分鐘，用玉米粉沾裹魚片，並用木棒敲打成薄片。

03　炸油燒至九分熱，先將魚頭及尾炸酥，再將魚一片片翻開下鍋，大火炸至熟且酥脆，撈出。

04　用 1 湯匙油爆香蔥、薑絲，放下綜合調味料煮滾，放入魚片，快速溜一下便盛盤。

Tips

桔子汁也可用鳳梨汁，或檸檬汁等代替。

INGREDIENTS •------

1 Fish, 1tbsp. Green onion shreds, 1tbsp. Ginger shreds, ½ cup Cornstarch, a little of Flour.

SEASONINGS

❶ ⅓ tsp. Salt, ½tbsp. Wine, a pinch of Pepper, 1tbsp. Egg white, 1tsp. Cornstarch.

❷ 2tbsp. Sugar, 1tbsp. Vinegar, 2tbsp. Water, 1tbsp. Soy sauce, ¼tsp. Salt, 3tbsp. Orange juice, ½tbsp. Cornstarch, 1tsp. Sesame oil.

STEP BY STEP •------

01　Cut off head and tail from fish, marinate with salt & wine, coat with flour and deep fry in heated oil to golden brown, put on two sides of a plate.

02　Debone the fish spine, slice the fish (skin-side down), every two slice of meat connected by the skin, so it looks like the wings of a butterfly. Marinate with ❶ for 10 minutes, coat with cornstarch. Knock fish to flat with a mallet.

03　Deep fry the fish slices to golden brown, drain.

04　Heat 1tbsp. of oil to fry ginger, green onion, add ❷ sauce, bring to a boil, mix with fish quickly, remove to the plate.

Tips

The orange juice can be replaced by pineapple or lemon juice.

魚中有餘

Fish in Fish's Boat

使用材料					
鮮魚	1 條（約 25 公分長）		醃魚肉料		
魚肉	10 兩		鹽	½ 茶匙	
香菇	3 朵		酒	½ 湯匙	
熟筍片	½ 杯		太白粉	½ 湯匙	
青、紅椒片	各少許				
葱段、薑片	酌量		調味料		
			酒	1 茶匙	
醃魚料			鹽	¼ 茶匙	
鹽	½ 茶匙		胡椒粉、麻油	少許	
酒	½ 湯匙				
胡椒粉	少許				

01　整條魚洗淨後由背部貼著大骨，兩面劃開，取出大骨，魚肉及尾部保持完整。用醃魚料抹勻，醃 10 分鐘，再投入熱油中炸酥。

02　另一塊魚肉順紋切成厚片，用醃魚肉料醃 10 分鐘；香菇泡軟再切片。將 2 杯油燒至八分熱，放下魚片泡熟（5 秒），瀝出。

03　另用 1 湯匙油爆香蔥段及薑片，放下香菇、熟筍片、青、紅椒片及魚片，灑下調味料，大火拌炒均勻，淋少許熱油便可盛入已炸好之整條魚中間。

INGREDIENTS •

1 Fish (25cm long), 400g. Fish fillet, 3 pieces Black mushroom, ½ cup Bamboo shoot slices(cooked), a little of Green pepper, a little of Red pepper, a little of Green onion, a little of Ginger slices.

SEASONINGS

❶ ½tsp. Salt, ½tbsp. Wine, a pinch of Pepper.

❷ ½tsp Salt, ½tbsp. Wine, ½tbsp. Cornstarch.

❸ 1tsp. Wine, ¼tsp. Salt, a pinch of Pepper, a few drops of Sesame oil.

STEP BY STEP •

01　Trim the fish, split the fish meat and debone the spine, make the fish meat into a large flat piece with the head and tail still attached. Marinate with seasonings ❶ for 10 minutes. Deep fry in heated oil to golden brown, drain. Then put on a plate as a fish boat.

02　Slice the fish fillet, marinate with seasonings ❷ for 10 minutes. Deep fry in 2 cups of 160°C oil for 5 seconds till done, drain. Soak black mushrooms to soft and slice it.

03　Heat 1tbsp. of oil to stir fry green onion and ginger, add all other ingredient slices and fish fillet, add seasonings ❸ and stir over high heat, mix evenly and place in the fish boat. Serve.

西炸鮭魚球

Deep-Fried Salmon Balls

使用材料

新鮮鮭魚	3 兩
馬鈴薯	1 斤
洋葱屑	½ 杯

調味料

鹽	½ 茶匙
胡椒粉	少許

西炸料

麵粉	3 湯匙
蛋	1 個
麵包粉	1 ½ 杯

01 馬鈴薯煮軟取出（馬鈴薯煮至筷子可插透即可），待稍涼後剝去外皮，用刀背壓成泥，放在大碗中。

02 鮭魚抹少許鹽後入鍋蒸熟，剝成小粒。用2湯匙油將洋蔥屑炒軟，連魚肉一起放入大碗中，加調味料仔細和薯泥拌勻，再分成小粒，並搓成橢圓形。

03 鮭魚球先沾一層麵粉，再沾上蛋汁，最後滾滿麵包粉，投入熱油中炸黃，瀝乾油裝入盤中，可另附美乃滋沾食。

INGREDIENTS

120g. Fresh salmon, 600g. Potato, ½ cup Onion, chopped.

SEASONINGS

❶ ½tsp. Salt, a pinch of Pepper.
❷ 3tbsp. Flour, 1 Egg, 1 ½cups Bread crumbs.

STEP BY STEP

01 Boil potato, peel and smash. Place in a large bowl.

02 Rub some salt on salmon, steam to done & tear into small pieces. Heat 2tbsp. of oil to stir fry onion, add salmon & ❶, mix well with mashed potato, divide into small oval balls.

03 Sprinkle flour on salmon balls, dip in beaten egg then coat with bread crumbs. Deep fry in heated oil to golden brown, drain and serve. You may serve with mayonnaise.

黃魚酥方

Deep-Fried Fish Cakes

使用材料

豆腐衣		6 張
黃魚		1 條 (約 12 兩)
		或白色魚肉 6 兩
荸薺		6 粒
葱屑		½ 杯
火腿屑		2 湯匙
土司麵包		6 片
甜麵醬		2 湯匙
葱段		酌量
牙籤		數支

醃魚料

蛋白	半個
鹽	¼ 茶匙
酒	少許
胡椒粉	少許

蛋麵糊

蛋	1 個
麵粉	⅓ 杯
水	⅔ 杯
鹽	少許

01　剔下黃魚肉或白色魚肉切成粗絲，用醃魚料醃一會兒。

02　荸薺切絲、擠乾，和蔥屑、火腿屑一起加入魚肉中拌勻。

03　豆腐衣裁去兩邊成長方形，在第 1 張塗一層蛋麵糊，再鋪上 1 張豆腐衣，然後塗糊並撒上魚肉料，另覆蓋 2 張豆腐衣及撒魚肉料，再蓋 2 張豆腐衣，用牙籤別住四邊（每張都要塗麵糊）。

04　放入熱油中，小火慢炸至酥，切成長方小塊排盤，附活頁土司麵包上桌，與蔥段、沾醬夾食（1 片土司先對切為 2，再片開成活頁狀。）

Tips

甜麵醬 1 湯匙加糖 ½ 湯匙、水 1 湯匙調稀，用少許油炒香。

INGREDIENTS •

6 pieces Bean curd sheet (dried), 1 Yellow croaker or 250g. fish fillet, 6 pieces Water chestnuts, ½ cup Green onion, chopped, 2tbsp. Ham, chopped, 6 slices Toast, 2tbsp. Sweet soybean paste, a little of Green onion, sectioned, Tooth picks.

SEASONINGS

❶ 1tbsp. Egg white, ¼tsp. Salt, some Wine, a pinch of Pepper.

❷ 1 Egg, ⅓ cup Flour, ⅔ cup Water, a pinch of Salt.

STEP BY STEP •

01　Take meat from yellow croaker, cut into shreds, marinate with ❶ for a while.

02　Shred the water chestnuts, mix with fish together with green onion and ham.

03　Brush with mixed ❷ on a dried bean curd sheet, cover with another sheet, Brush with ❷ again, add some fish mixture and spread, cover with 2 slices bean curd sheet and spread fish mixture, cover with 2 more bean curd sheets seal with tooth picks. Brush with ❷ between every piece of dried bean curd sheet

04　Deep fry in heated oil(about 160°C) over low heat to golden brown. Cut into rectangles, serve with toast, sweet soybean sauce and green onion sections.

Tips

Mix 1tbsp. sweet soybean paste with ½tbsp. sugar, 1tbsp. water, then stir fry with 1tbsp. heated oil. This is the sauce.

香菇肉燥蒸鮭魚

Steamed Salmon with Meat Sauce

01　將鮭魚切成 3 公分寬、5 公分長、半公分厚的長方片。嫩豆腐也切成相同大小，兩者相間隔鋪排在深盤中。

02　將 2 湯匙肉燥撒在鮭魚及豆腐上，入鍋蒸 10 分鐘端出，撒下蔥屑，淋上 1 湯匙熱油即可上桌。

 Tips

‣ 除肉燥外還可以用炒過的蝦籽（蝦膏）或肉醬來蒸。

‣ 用油炒香絞肉及香菇屑，加酒、醬油、糖、鹽、五香粉、紅蔥酥及水同煮至肉香汁乾便是肉燥（約 1 小時）。

INGREDIENTS ◆

150g. Salmon, 1 pack Bean curd, 2tbsp. Meat sauce, 1tbsp. Green onion, chopped.

STEP BY STEP ◆

01　Cut salmon into 3cm x 5cm x 0.5cm slices. Slices bean curd into same size. Place fish and bean curd slice one against the other on a plate.

02　Pour 2tbsp. meat sauce on top of fish, steam for 10 minutes. Sprinkle green onion and splash 1tbsp. heated oil. Serve.

 Tips

· Fried dried-Shrimp-roe can substitute the meat sauce.

· Stir fry minced meat & black mushroom cubes, add wine, soy sauce, sugar, salt, five spicy powder, dried red scallion and water, cook over low heat for about 1 hour till sauce reduced and becomes fragrant. This is the meat sauce.

鱸魚雙味

Two Ways Fish

使用材料

新鮮鱸魚（或其他新鮮魚）	1 條		酒	1 湯匙
	（約 1 ½ 公斤）		淡色醬油	1 湯匙
鹽	少許			
太白粉	2 茶匙		**蒸魚料**	
五香花椒鹽	少許		豆豉	1 湯匙
			紅辣椒屑	½ 湯匙
醃魚料			火腿屑	1 湯匙
鹽	¼ 茶匙		薑屑	½ 茶匙
蔥	1 支		蔥花	1 湯匙
薑	1 片			

01 切下魚頭及魚尾，用少許鹽和酒抹一下，裹上麵粉。再將半邊魚肉切下來。魚肉斜刀切成大片，用醃魚料醃 10 分鐘以上。

02 用熱油炸酥魚的頭和尾後，排在盤子的兩端。再將魚肉下鍋也炸酥，撈出後灑上少許五香花椒鹽，排另一邊。

03 另一邊的魚肉也切下來，切成大片，拌少許太白粉及鹽，放在蒸盤內，撒上蒸魚料，大火蒸 10 分鐘，取出後排在另一邊（中間排一列檸檬片間隔）。

此係椒鹽魚片與豆豉辣椒蒸魚兩種口味，也可改用酥炸魚捲、蠔油魚片或麵拖魚條等其他口味之魚的菜式。

INGREDIENTS •

1 Fish (about 900g), a pinch of salt, a pinch of Five spice powder with brown pepper corn salt, 2tbsp. Cornstarch.

SEASONINGS

❶ ¼tbsp. salt,1 Ginger slice, 1 Stalk green onion, 1tbsp. Wine, 1tbsp. light colored Soy sauce.

❷ 1tbsp. Fermented black bean, ½tbsp. Red chili (chopped), 1tbsp. Ham (chopped), ½tsp. Ginger (chopped), 1tbsp. Green onion (chopped).

STEP BY STEP •

01 Cut off head and tail from fish, marinate with salt and wine, coat with flour. Remove one side of fish meat from the spine, slice and marinate with seasonings ❶ for 10 minutes.

02 Heat the oil to deep fry head and tail to golden brown. Put on two sides of a plate. Also deep fry fish meet till golden brown. Arrange on one side of the plate, salt.

03 Slice another half fish meat, mix with a little of salt and cornstarch, arrange on a plate , pour ❷ on, steam for 10 minutes. Arrange on the other side of the plate.(Divided by lemon slices as decoration).

The flavor can be changed to oyster flavor or deep-fried fish rolls.

雙菇燜鮮魚

Fish with Mushrooms

01 將魚頭及尾切下，由背部剖開，取下兩邊之魚肉。魚頭、魚尾和魚骨沾上乾麵粉，用熱油炸黃，和盤中墊底用。

02 魚肉切 1 ½ 寸大小，用醃魚料拌醃，約 5 分鐘。沾上一層乾麵粉後再裹上蛋糊，炸黃。

03 用 1 湯匙熱油爆香蔥段、薑片，淋下高湯（或水），放下魚塊及調味料，小火煮 1 分鐘。

04 撈出魚塊堆放在魚骨上，湯汁中加入香菇絲（泡軟切絲）和金菇（切段），大火煮滾片刻，勾成薄芡，淋到魚面上即成。

INGREDIENTS •

1 Grouper (or other fish), 3 Black mushrooms, 100g. Needle mushrooms, 1stalk Green onion, Ginger slices, 1 cup Soup stock, 4tbsp. Flour(for coat the fish).

SEASONINGS

❶ ½tbsp. Wine, ¼tsp. Salt, a pinch of Pepper.

❷ 1 Egg, 4tbsp. Flour, Water (moderate amount).

❸ ½tbsp. Soy sauce, ½tsp. Salt.

STEP BY STEP •

01 Cut off head & tail from fish, debone fish spine, leave fish meat for later use. Coat fish head, tail and spine with flour, deep fry in heated oil to golden brown. Arrange on a plate.

02 Cut fish meat into 1 ½" long pieces, marinate with seasonings ❶ for 5 minutes, coat with flour and dip in ❷ batter, then deep-fry in heated oil to golden brown.

03 Heat 1tbsp. of oil to stir fry green onion & ginger, add soup stock, ❸ and fish, cook over- low heat for 1 minute.

04 Take fish out from wok and arrange on top of bones. Add soaked black mushroom (shredded) & needle mushroom. Bring to a boil, thickened and pour over fish.

四味芝麻魚

Four Flavors Fish

使用材料

新鮮鱸魚	1 條（或魚肉 1 斤）
麵粉	3 湯匙
蛋	1 個
白芝麻	½ 杯

醃魚料

蔥	1 支
薑	2 片
鹽	½ 茶匙
胡椒粉	¼ 茶匙

四味料

美乃滋、蕃茄醬、五香花椒鹽 各酌量

蠔油醬 （油 ½ 湯匙、蠔油 1 湯匙、水 2 湯匙煮滾）

01 取下兩邊之魚肉後，將魚頭及魚尾切下，沾上麵粉炸黃，排在大盤中。

02 魚肉去皮，順紋切成拇指般粗條，用醃魚料拌醃片刻。

03 魚條上撒少許麵粉，再沾上蛋汁，最後沾裹上白芝麻，全部沾好，投入七分熱的溫油中，小火慢炸至熟，撈出瀝乾油。

04 將魚條排列在魚骨上，並將四小碟不同之調味料附上，以供沾食。

INGREDIENTS •

1 Fresh fish, 3tbsp. Flour, 1 Egg, ½ cup Sesame seeds.

SEASONINGS

❶ 1 stalk Green onion, 2 Ginger slices, ½tsp. Salt, ¼tsp. Pepper.

❷ Mayonnaise.

❸ Ketchup.

❹ Five spice powder with brown pepper corn salt.

❺ Oyster sauce : ½tbsp. oil, 1tbsp. Oyster sauce, 2tbsp. Water, bring to a boil.

STEP BY STEP •

01 Cut off head & tail from fish, coat with flour and deep fry in heated oil to golden brown. Arrange on the plate.

02 Cut the fish meat into thumb size sticks, marinate with ❶ for a while.

03 Sprinkle flour on fish, dip in beaten egg and coat with sesame seeds. Deep fry with 150°C oil to golden brown. Drain. Arrange on top of fish head.

04 Serve with ❷❸❹❺ four kinds of dipping sauce.

脆皮五柳魚

Crispy Fish with Assorted Strings

使用材料

新鮮魚 .... 1 條

洋葱絲、青椒絲、胡蘿蔔絲、金菇
絲、木耳絲 .... 各酌量

蛋麵糊

蛋 .... 1 個

麵粉 .... 2 大匙

太白粉 .... 2 大匙

鹽 .... ¼ 茶匙

油 .... 2 茶匙

水 .... 適量

調味料

鹽 .... ¼ 茶匙

淡色醬油 .... 1 湯匙

糖 .... 1 湯匙

醋 .... 1 湯匙

水 .... 4 湯匙

麻油、胡椒粉 .... 各少許

太白粉水 .... 酌量

醃魚料

鹽 .... ¼ 茶匙

胡椒粉 .... ¼ 茶匙

01 魚清理乾淨之後，由腹部下刀片切開，但背部仍需連著，剪除大骨，撒下鹽及胡椒粉醃片刻。

02 沾上蛋麵糊，用熱油炸脆後撈出，放入盤中。

03 用 1 湯匙油炒香洋葱絲及胡蘿蔔絲，加入鹽、淡色醬油、糖、醋及水調味，煮滾後，放下青椒、金菇及木耳絲，再用太白粉水勾芡，淋下麻油及胡椒粉，全部淋到魚上即可。

INGREDIENTS •

1 Fish (any kind), Onion, Green pepper, Carrot, fungus, Needle mushroom, each a little amount (shredded).

BATTER

❶ 1 Egg, 2tbsp. Flour, 2tbsp. Cornstarch, ¼tsp. Salt, 2tsp. Oil, suitable amount of Water.

SEASONINGS

❷ ¼tsp. Salt, 1tbsp. Light colored soy sauce, 1tbsp. Sugar, 1tbsp. Vinegar, 4tbsp. Water, a few drops of Sesame oil, a pinch of Pepper, some Cornstarch paste.

❸ ¼tsp. Salt, ¼tsp. Pepper.

STEP BY STEP •

01 Trim the fish. Slice from the belly part, remove the spine. Marinate with salt & pepper for a while.

02 Sprinkle flour on fish, coat with batter, deep fry in heated oil till golden brown, drain and put on a plate.

03 Heat 1tbsp. of oil to stir fry onion & carrot, add seasonings ❷, bring to a boil, add green pepper, needle mushroom and soaked fungus, then thickened with Cornstarch paste and pour over fish. Serve.

銀紙烤鮭魚

Baked Salmon in Silver Package

使用材料				
	新鮮鮭魚	12 兩	淡色醬油	2 湯匙
	香菇	2 朵	糖	½ 茶匙
	青椒	¼ 個	胡椒粉	少許
	葱絲	1 湯匙	水	2 湯匙
	鋁箔紙	2 張（6寸四方）		
	檸檬片	3 片		

醃魚料

酒	1 湯匙

01 　新鮮鮭魚切成 4 片，用醃魚料拌醃 10 分鐘。

02 　香菇泡軟切絲；青椒也切絲。

03 　鋁箔紙上塗上油後，放入魚肉，再將香菇、青椒、蔥絲，分別撒在魚肉上，
淋下醃魚料，折角包妥。

04 　放入烤箱中大火烤熟（約 8 ～ 10 分鐘），附檸檬片上桌。

 Tips

▸ 也可 1 片鮭魚包成 1 包，共包成 4 小包。

▸ 可用平底鍋採蒸烤方式：平底鍋中放水 1 杯，放下鋁箔包，蓋上鍋蓋，約 10 分鐘。

INGREDIENTS ·

300g. Fresh salmon fillet, 2 Black mushrooms, ¼ Green pepper, 1tbsp. Green onion shreds, 2 pieces Aluminum foil (6"x6"), 3 pieces Lemon.

SEASONINGS

❶ 1tbsp. Wine, 2tbsp. light colored Soy sauce, ½tsp. Sugar,a pinch of Pepper, 2tbsp Water.

STEP BY STEP ·

01 　Cut salmon to 4 big pieces Marinate salmon with ❶ for 10 minutes.

02 　Soak black mushroom to soft and slice it. Shred the green pepper.

03 　Rub some oil on center of aluminum foil, put the fish on, arrange black mushroom, green pepper and green onion on fish, then add seasonings ❶ , fold and wrap into a package.

04 　Bake in oven over high heat for 8~10 minutes. Serve with lemon slices.

 Tips

· You may pack salmon into 4 small packages.
· You may use frying pan to steam it. Add 1 cup of water in pan, put salmon package on, cover and steam for 10 minutes.

家常海鰻絲

Shredded Eel, Home Style

<table>
</table>

使用材料	海鰻	半斤	胡椒粉	少許
	芹菜段、木耳絲	各半杯	太白粉	2茶匙
	熟筍絲	⅓杯		
	紅辣椒絲	1湯匙	**調味料**	
	葱、薑絲	各少許	辣豆瓣醬	½湯匙
	太白粉	½杯	醬油	2茶匙
			糖	1茶匙
	醃魚料		酒	1茶匙
	酒	1茶匙	清水	⅓杯
	蛋白	半個	麻油	½茶匙
	鹽	¼茶匙	醋	1茶匙

01　鰻魚去除大骨及魚皮後，順紋切成粗絲，全部用醃魚料拌勻醃 10 分鐘，再裹上一層太白粉。用熱油將海鰻絲炸酥，撈起。

02　起油鍋，用 2 湯匙油炒香蔥、薑絲，加入辣豆瓣醬同炒，並加入醬油、糖、酒及水，放入鰻魚絲、木耳絲及筍絲拌炒，燒煮一下，見汁將吸乾，放下芹菜段、紅辣椒絲與麻油，沿鍋邊淋醋，拌勻即可裝盤。

INGREDIENTS •

300g. Eel, ½ cup Celery sections, ½ cup fungus shreds, ⅓ cup Bamboo shoot shreds, 1tbsp. Red chili, shreds, a little of Green onion shreds, a little of Ginger shreds, ½ cup Cornstarch.

SEASONINGS

❶ 1tsp. Wine,1tbsp. Egg white, ¼tsp. Salt, a pinch of Pepper, 2tsp. Cornstarch.

❷ ½tbsp. Hot bean paste, 2tsp. Soy sauce, 1tsp. Sugar, 1tsp. Wine, ⅓ cups Water, ½tsp. Sesame oil, 1tsp. Vinegar.

STEP BY STEP •

01　Debone the eel and remove the skin, cut into strings along the grain, marinate with ❶ for 10 minutes. Coat with cornstarch, deep fry in heated oil to golden brown.

02　Heat 2tbsp. of oil to stir fry green onion, ginger, add hot bean paste and seasonings ❷ , then add eel, fungus and bamboo shoot, cook over low heat until liquid dissolved. Add celery, red chili, sesame oil, finally, add vinegar, mix well. Remove to a plate. Serve.

蟹肉荷包豆腐

Bean Curd Balls Stuffed with Crab

使用材料				
	豆腐	3 塊	太白粉	1 湯匙
	蟹肉	½ 杯	胡椒粉	少許
	蔥末	1 湯匙		
	蔥段	5 小段	**調味料**	
	薑片	3 小片	酒	1 茶匙
	太白粉	適量	高湯	1 ½ 杯
	青菜	隨意	醬油	1 ½ 湯匙
			糖	½ 茶匙
	豆腐料		鹽	¼ 茶匙
	鹽	½ 茶匙	胡椒粉	少許
	蛋白	1 個	太白粉水	少許

01 將豆腐壓成泥，加入豆腐料拌勻。

02 爆香少許蔥末，將蟹肉炒過。

03 湯匙中塗上油，將豆腐餡料鋪在湯匙上，放少許蟹肉再加蓋一些豆腐餡，做成橢圓形，全部蒸熟（約 5 ～ 6 分鐘），取下豆腐片，沾上一層太白粉，用熱油炸黃。

04 用 1 湯匙油炒香蔥段、薑片，淋下酒及高湯等調味料（除太白粉水），調味後放入豆腐同煮，約 3 分鐘後即可勾芡盛盤，用少許青菜裝飾，即可享用。

INGREDIENTS •

3 pieces Bean curd, ½ cup Crab meat, 1tbsp. Green onion, chopped, 5 pieces Green onion sections, 3 pieces Ginger slice, suitable amount of Cornstarch(for coating), Green vegetables.

SEASONINGS

❶ ½tsp. Salt, 2 tbsp. Egg white, 1tbsp. Cornstarch, a pinch of Pepper.

❷ 1tsp. Wine, 1 ½cups Soup stock, 1 ½tbsp. soy sauce, ½tsp. Sugar, ¼tsp. Salt, a pinch of Pepper, some Cornstarch paste.

STEP BY STEP •

01 Smash bean curd and mix with ❶ .

02 Heat 2tbsp. of oil to stir fry chopped green onion then add crab meat in, mix thoroughly.

03 Brush oil on the spoons, put ½tbsp. bean curd mixture on a spoon, stuff 1tsp. crab meat in, then cover with ½tbsp. bean curd mixture, make oval balls and steam for 5~6 minutes. Remove bean curd, coat with cornstarch. Deep fry in heated oil to golden brown.

04 Heat 1tbsp. of oil to stir fry green onion sections and ginger, add wine and soup stock, add seasonings and cook with bean curd for 3 minutes. Thicken with cornstarch paste, put on a plate, decorate with green vegetables. Serve.

酥皮豆腐捲

Crispy Bean Curd Rolls

使用材料

老豆腐	2 方塊
豆腐衣	2 張
鹽	½ 茶匙

麵糊

麵粉	1 湯匙
水	2 湯匙

01　將豆腐直切成 1 公分厚而長的片狀，兩面均抹上鹽，再用少量熱油煎黃。

02　豆腐衣 1 張切成長方形，先放上一長片豆腐捲一圈，再放第二片再捲，待第三片放上並捲完後，用麵糊黏妥封口處，做好 2 條，投入熱油中去炸。

03　待豆腐捲炸至金黃色而外皮酥脆即可撈出，趁熱切成小塊，附花椒鹽或辣醬油沾食。

INGREDIENTS ·

2 pieces Bean Curd, 2 pieces Dried bean curd sheet, ½tsp. Salt.

FLOUR PASTE

1tbsp. Flour, 2tbsp. Water.

STEP BY STEP ·

01　Cut bean curd into thick slices, sprinkle salt, fry with 3tbsp. of hot oil until browned.

02　Cut bean curd sheet into rectangle, place the bean curd on, fold and roll once, place another bean curd on, fold and roll again, place the third layer of bean curd, fold and roll again, finally stick the edge with flour paste. Deep fry in hot oil to golden brown.

03　Cut the bean curd roll into pieces, serve with on Worcester sauce or brown pepper corn powder with salt .

八寶豆腐盒

Bean Curd Treasure Boxes

板豆腐	3 方塊	
蝦仁	3 兩	
香菇	3 朵	
荸薺	5 粒	
薑屑	1 茶匙	
麵糊	少許	

拌蝦料

葱屑	1 湯匙
鹽	¼ 茶匙

蛋白	½ 個
太白粉	1 茶匙

調味料

酒	1 茶匙
醬油	1 湯匙
高湯（或水）	1 杯
太白粉水	酌量
葱花	少許

01 豆腐每長方塊切為二，用熱油大火炸透，撈出後，在豆腐⅓厚的部分切一刀口，使成活頁狀，挖出中間少許豆腐。

02 蝦仁、香菇、荸薺切小丁，加入拌蝦料，填入豆腐中，刷少許麵糊，蓋上蓋子。

03 用 2 湯匙油來煎豆腐盒（蓋朝下），加入薑屑及調味料，小火煮 5 分鐘，盛出豆腐盒。湯汁勾芡，撒少許蔥花，淋到豆腐盒上。

Tips

這道菜也可將餡料炒熟，以盤盛裝上桌，用剛炸透之豆腐自行填塞而食之。

INGREDIENTS •

3 pieces Bean curd (2" x 2"), 120g. Shrimp, shelled, 3 pieces Black mushrooms, 5 pieces Water chestnut, 1tsp. Ginger, chopped, a little of Flour paste.

SEASONINGS

❶ 1tbsp. Green onion, chopped, ¼tsp. Salt, 1tbsp. Egg white, 1tsp. Cornstarch.

❷ 1tsp. Wine, 1tbsp. Soy sauce, 1cup Soup stock, suitable amount of Cornstarch paste, 1tbsp. Green onion, chopped.

STEP BY STEP •

01 Cut bean curd into 2 squares, deep fry in heated oil to golden brown. Remove and cut a ⅓ " thick slice as a cover lid (do not cut through), scoop out some bean curd.

02 Cut shrimp, black mushroom and water chestnuts to small pieces, mix with seasoning ❶ , stuff in bean curd box. Cover the lid and seal with flour paste.

03 Heat 2tbsp. of oil to fry bean curd box with lid side down. Add in ginger and seasoning ❷ , cook over low heat for 5 minutes. Remove bean curd box out and put on a plate, thicken the soup, pour over the bean curd, sprinkle green onion.

Tips

The stuffing can be stir-fried and stuff directly in bean curd while eating.

豆腐咕咾肉

Sweet and sour Bean Curd

使用材料	嫩豆腐	2 方塊	糖、醋	各 2 湯匙
	鹽	1 茶匙	鹽	½ 茶匙
	青椒	1 個	太白粉	2 茶匙
	洋葱	半個	水	5 湯匙
	荔枝（或紅毛丹）	10 粒	麻油	1 茶匙
	（可用罐頭或免用）			
	鳳梨丁	半杯		
	拌雞料			
	番茄醬	2 湯匙		

01 將豆腐切成小方塊後，撒下鹽 1 茶匙，放置 1 分鐘後，拭乾水分，投入熱油，中大火炸至外皮脆硬為止。

02 青椒去籽、切片；洋蔥切丁備用。

03 加入 2 湯匙油，起油鍋炒香洋蔥丁，加入番茄醬炒紅，再加入青椒片及水果同炒，淋下綜合調味料煮滾，倒入豆腐，快速拌合即可盛出。

INGREDIENTS •

2 pieces Bean curd, 1tsp. Salt, 1 Green pepper, ½ Onion, 10 Li-Chi (canned), ½ cup Pineapple.

SEASONINGS

2tbsp. Ketchup, 2tbsp. Sugar, 2tbsp. Vinegar, ½tsp. Salt, 2tsp. Cornstarch, 5tbsp. Water, 1tsp. Sesame oil.

STEP BY STEP •

01 Cut bean curd into cubes, sprinkle 1tsp. salt and leave for 1 minute. Pat dry. Deep fry in heated oil to golden brown.

02 Remove seeds from green pepper, cut green pepper and onion into cubes.

03 Heat 2tbsp. of oil to stir fry onion and ketchup, then add in pepper and fruits, add in seasonings, bring to a boil, mix with bean curd. Arrange on a plate. Serve.

香酥豆腐鬆

Crispy Minced Bean Curd

老豆腐	3 方塊
香菇	2 朵
榨菜丁	2 湯匙
荸薺 4 粒或筍丁	半杯
熟胡蘿蔔丁	2 湯匙
油條	1 根
核桃屑	1 湯匙
生菜葉	12 枚
香菜	少許

綜合調味料

醬油	2 湯匙
鹽、糖	各 ½ 茶匙
胡椒粉	少許
水	3 湯匙
太白粉、麻油	各少許

01 將豆腐切成厚片，先在油中略煎，再切成小丁，其他材料均切成小丁，油條切薄片在油中炸酥（或用烤箱烤脆），堆放在盤中。

02 加入 2 湯匙油，起油鍋炒香香菇，加入豆腐炒乾，再放榨菜、荸薺及熟胡蘿蔔丁同炒，並淋下綜合調味料拌炒均勻，裝盤。

03 將香菜屑及核桃屑撒在豆腐鬆上，隨生菜上桌包食之。

INGREDIENTS •

3 pieces Bean curd (2" x 2"), 2 pieces Black mushroom, 2tbsp. Pickled kale head, 4 pieces Water chestnut, 2tbsp. Carrot, cooked, 1 Yiou-Tiau, 1tbsp. Walnut, chopped, 12 pieces Lettuce leaf, a pinch of Chinese parsley.

SEASONINGS

2tbsp. Soy sauce, ½tsp. Salt, ½tsp. Sugar, a pinch of Pepper, 3tbsp. Water, some Cornstarch, a few drops of Sesame oil.

STEP BY STEP •

01 Slice the bean curd, fry to browned, cut into small cubes. Cut all other ingredients into small cubes, cut Yiou-Tiau into 1" sections, deep fry or bake to crispy, put on a plate.

02 Heat 2tbsp. of oil to stir fry black mushroom, bean curd, kale head, water chestnuts and carrot, add in seasonings, stir evenly, semove to a plate.

03 Sprinkle Chinese parsley and walnut on top. Serve with lettuce leaves.

油淋黃雀

Fried Crispy Packages

使用材料

豆腐衣	5 張
絞肉	3 兩
香菇	4 朵
青江菜	半斤
葱、薑屑	各少許

調味料

醬油	2 湯匙
鹽、胡椒粉	各少許

油淋料

麻油	1 ½ 湯匙
葱花	1 湯匙
鹽	¼ 茶匙
花椒粉	少許

01 香菇泡軟，切碎；青江菜燙軟沖涼後剁碎，擠乾水分。

02 用 2 湯匙油炒香蔥、薑屑及絞肉，加入香菇、青江菜續炒，調味後盛出。

03 豆腐衣一切為二，包捲②項料，從尖角包起，先捲成長條，再打一個結成黃雀狀，用油炸酥。

04 炸好之黃雀放大碗中，撒下蔥花、鹽和花椒粉，淋下燒熱之麻油，抖動大碗，拌勻、裝盤。

INGREDIENTS •

5 pieces Dried bean curd sheet, 120g. Minced pork, 4 pieces Black mushroom, 300g. Green cabbage, 1tsp. Green onion, chopped, 1tsp. Ginger, chopped.

SEASONINGS

❶ 2tbsp. Soy sauce, a pinch of Salt, a pinch of Pepper.

❷ 1 ½tbsp. Sesame oil, 1tbsp. Green onion, chopped, ¼tsp. Salt, a pinch of Brown pepper corn powder.

STEP BY STEP •

01 Chop soaked black mushroom. Blanch green cabbage, rinse, chop and squeeze dry.

02 Heat 2tbsp. of oil to stir fry green onion, ginger, minced pork, add in black mushroom, green cabbage. Season with ❶ .

03 Cut each bean curd sheet into 2 pieces, wrap item ② material, roll into a cylinder and tie two ends into a knot. Deep fry to crispy.

04 Put on a plate, sprinkle green onion, salt, brown pepper corn powder, splash heated sesame oil, mix and serve.

炸百花蛋

Deep Fried Stuffed Egg

使用材料			拌蝦料	
新鮮雞蛋	6 個		蛋白	½ 個
鹽	少許		鹽	½ 茶匙
太白粉	少許		酒	1 茶匙
蝦仁	6 兩		太白粉	½ 湯匙
荸薺	5 個		麻油	少許
米粉或芥蘭菜葉	酌量			

01 鍋中放冷水煮蛋，用筷子轉動蛋，以使蛋黃熟後的位置保持在中間部分。蛋煮熟後剝殼，每個對切兩半，撒少許鹽放片刻。

02 蝦仁壓成泥狀，加入切碎並擠乾水分的荸薺，用拌蝦料仔細拌勻，成為有彈性之蝦餡。

03 灑少許太白粉在蛋面上，再放 1 湯匙的蝦餡，塗成半凸形狀，沾上一層太白粉。

04 將炸油燒至七分熱，放下③項的百花蛋，用慢火炸熟蝦面，使成為金黃色為止，每個橫切兩半，排在炸鬆米粉或芥蘭菜葉上即可。

INGREDIENTS

6 Eggs, a pinch of Salt, some Cornstarch, 240g. Shrimp, shelled, 5 pieces Water chestnuts, 5 pieces Mustard green leaf.

SEASONINGS

1tbsp. Egg white, ½tsp. Salt, 1tsp. Wine, ½tbsp. Cornstarch, a few drops of Sesame oil.

STEP BY STEP

01 Put egg in a wok with cold water, boil and stir the eggs with chopsticks tenderly to make every egg yolk in the accurate center. When the egg is done, shelled, cut into halves, sprinkle salt.

02 Smash the shrimp, mix with crushed water chestnuts and seasonings, stir to sticky.

03 Sprinkle cornstarch on the cutting surface of eggs, put 1tbsp. shrimp mixture to form a ball egg, coat with cornstarch.

04 Deep fry in heated oil over low heat till the shrimp is done. Cut each egg ball into halves (Cut from the middle line of each egg). Decorate with fried vegetable shreds. Serve.

蠔油蛤蜊蛋

Steamed Egg with Clams

使用材料				
	雞蛋	4 個	鹽	¼ 茶匙
	冷高湯或水	2 杯	胡椒粉	少許
	蛤蜊	15 粒	蠔油	1 湯匙
	葱	1 支	太白粉水	酌量
	薑	1 片		
	香菜	少許		
	調味料			
	酒	1 茶匙		

01 蛋加鹽少許，打散，加入清水調勻，過濾到深盤中，入鍋以小火蒸至熟。

02 蛤蜊用清水 1 ¼ 杯煮至殼微開即撈出，剝肉，汁留用。

03 用 1 湯匙油煎香葱支和薑片，淋酒及蛤蜊湯，煮滾後調味並勾芡，加入蛤蜊肉，再一滾即可淋到蛋面上，可撒下少許香菜或紅葱酥。

INGREDIENTS •

4 Eggs, 2cups Soup stock, 15 pieces Clams, 1 stalk Green onion, 1 Ginger slice, some Chinese parsley.

SEASONINGS

1tsp. Wine, ¼tsp. Salt, a pinch of Pepper, 1tbsp. Oyster sauce, some Cornstarch paste.

STEP BY STEP •

01 Beat the eggs, mix with salt and water, sieve into a deep plate, steam over low heat until done.

02 Cook clams with 1 ¼ cups of water, remove clams meat. Leave the soup for later use.

03 Heat 1tbsp. of oil to fry green onion & ginger, add wine and clams soup, bring to a boil, thicken with cornstarch paste and add clams. Pour over steamed egg, sprinkle the Chinese parsley.

釀百花素雞

Stuffed Shrimp in Bean Curd Cake

使用材料		
蝦仁	4 兩	
絞肥肉	少許	
大素雞	2 條	
西洋菜或小芥蘭菜	數支	

拌蝦料

蛋白	半個
蔥薑水	2 湯匙
鹽	¼ 茶匙
麻油	½ 茶匙
太白粉	1 湯匙

調味料

高湯	1 杯
蠔油	1 ½ 湯匙
太白粉水	1 茶匙
麻油	少許

01　蝦仁洗淨擦乾,壓成泥狀。放在大碗中,加入絞肥肉及拌蝦料,仔細拌勻。

02　素雞切成雙飛(活頁)厚片,切口抹少許乾太白粉,夾入蝦泥料,用水將蝦泥抹光滑。

03　青菜切段在開水中燙軟,放在盤中,上面排放夾好之素雞,滾水大火蒸 8 分鐘。蒸好後湯汁泌到鍋中,加高湯及調味料,勾芡再淋到百花素雞上。

INGREDIENTS •

150g. Shrimp, shelled, 1tbsp. Pork fat, minced, 2 pieces. Bean curd cake, Green vegetables.

SEASONINGS

❶ 1tbsp. Egg white, 2tbsp. Green onion & Ginger juice, ¼tsp. Salt, ½ tsp. Sesame oil, 1tbsp. Cornstarch.

❷ 1cup Soup stock, 1 ½tbsp. Oyster sauce, 1tsp. Cornstarch paste, a few drops of Sesame oil.

STEP BY STEP •

01　Clean the shrimp, then smash it. Mix with pork fat & seasonings ❶ .

02　Cut the bean curd cake into butterfly shapes, stuff inside with shrimp paste. (Sprinkle with a little bit of cornstarch first).

03　Blanch vegetables, put on a plate, arrange bean curd cake on, steam for 8 minutes over high heat. Boil seasoning ❷ . Pour over and then serve hot.

燻三絲齋鵝

Smoked Vegetarian Goose

使用材料	新鮮豆腐包	2 塊	高湯（泡香菇水）	½ 杯
	豆腐衣	4 張	麻油	½ 湯匙
	熟筍絲	½ 杯		
	香菇絲	½ 杯	燻料	
	熟胡蘿蔔絲	½ 杯	黃糖、米、茶葉	各⅓杯
	調味料			
	醬油	2 湯匙		
	糖	1 茶匙		

01　用 1 湯匙油炒香菇絲、熟筍絲及熟胡蘿蔔絲，放入調味料煮透後，瀝出菜料，湯汁留用。

02　豆腐衣 2 張相對平放，刷上①項湯汁，再將 1 塊新鮮豆腐包攤開放在上面，也刷上湯汁，再鋪上一半量之①項料，捲包成筒狀。做好 2 個齋鵝。

03　入鍋大火蒸 5 分鐘，取出待稍涼即可燻。

04　鍋中鋪 1 張鋁箔紙，上放燻料，燻架子塗油後放上齋鵝，先大火，待煙冒出後，改小火燻 5 分鐘，翻面再燻 3 分鐘，取出切塊上桌。

Tips

喜燻味重者，可燻久一點，燻的食物宜冷食較香。

INGREDIENTS •

2 pieces Fresh bean curd package, 4 pieces Dried bean curd sheet, ½ cup Cooked bamboo shoot shreds, ½ cup Black mushroom shreds, ½ cup Cooked carrot shreds.

SEASONINGS

2tbsp. Soy sauce, 1tsp. Sugar, ½ cup Soup stock, ½tbsp. Sesame oil.

SMOKE MATERIALS

⅓ cup Sugar, ⅓ cup Rice, ⅓ cup Tea leaves.

STEP BY STEP •

01　Heat 1tbsp. of oil to stir fry mushroom, bamboo shoot & carrot shreds, add seasonings, bring to a boil, drain. Leave the soup for later use.

02　Place 2 bean curd sheets opposite side, Brush the soup from ① , then place the fresh bean curd package on top. Put ½ vegetables in center, brush some soup again and roll into a cylinder. Make 2 rolls. This is the vegetarian goose.

03　Steam for 5 minutes. Take out and let cooled.

04　Put a piece of aluminum foil in a wok, put smoke materials on. Brush oil on a rack. Put goose on, cover the lid, smoke over medium heat for 5 minutes, turn goose over and smoke for 3 more minutes. Cut and serve after it cooled.

Tips

If you like it smell stronger, smoke longer.

金華扣四蔬

King-Hwa Ham & Vegetables

<table>
<tr><td>使用材料</td><td>中國火腿</td><td>10 片</td><td>鹽</td><td>少許</td></tr>
</table>

使用材料

中國火腿	10 片
白蘆筍罐頭	1 罐
青江菜或小芥蘭	10 支
草菇	20 粒
大白菜	半斤

調味料

| 高湯 | 2 ½ 杯 |

| 鹽 | 少許 |
| 太白粉水 | 酌量 |

01 中國火腿整塊蒸熟，切成 1 寸寬、2 寸長的薄片。青江菜取中間嫩的部分大約 2 寸長，用滾水（加少許鹽）燙煮一下，撈出，沖涼擠乾。草菇也燙一下，撈出。

02 大白菜切寬條，用高湯 1 杯煮 5 分鐘、瀝出。

03 選 1 只中型淺盒或較深的湯盤，先在中間排火腿片，左右兩邊分別排青江菜及白蘆筍，上下二端排列草菇，中間以大白菜填補凹處，淋入 1 杯高湯，大火蒸 5 分鐘。

04 將③項之湯汁泌入鍋內（太少時可再加半杯高湯），另將火腿等倒扣在大盤中。湯汁加少許鹽調味後，用太白粉水勾芡，淋少許雞油，淋在四蔬上即可上桌。

INGREDIENTS •

10 pieces Chinese King-Hwa ham, 1 can Asparagus, 10 pieces green cabbage, 20 pieces straw mushroom, 300g. Chinese cabbage.

SEASONINGS

2 ½cups Soup stock, a pinch of Salt, suitable amount of Cornstarch paste.

STEP BY STEP •

01 Steam the Chinese King-Hwa ham to done, cut into thin slices (1" x 2"). Trim the green cabbage (using the heart part). Blanch in boiling water, season with salt, drain and soak in cold water, squeeze dry. Blanch the straw mushrooms in boiling water, too.

02 Cut Chinese cabbage to thick sticks, cook with 1cup of soup stock for 5 minutes, drain.

03 Choose a deep plate or bowl, arrange ham slices in the center of the bowl, put asparagus and green cabbage on two sides of ham, then put straw mushroom on the left space. Stuff center with Chinese cabbage. Add 1cup of soup stock, steam for 5 minutes.

04 Pour the juice from steamed bowl, add another ½ cup of soup stock. Reverse the plate to let the ham & vegetable stand on a large plate. Boil the steamed soup, season with salt, thicken with cornstarch paste, sprinkle some chicken grease ,pour over the vegetables. Serve.

蒜汁肉捲蘆筍

Asparagus in Bacon Rolls

使用材料		
綠蘆筍	15 支	
培根	15 片	
調味汁		
大蒜泥	½ 湯匙	
水	1 ½ 湯匙	
甜醬油	1 ½ 湯匙	
紅油、麻油	各 1 茶匙	

01　綠蘆筍削去老皮後切為二段，在滾水中燙熟（水中加鹽）撈出沖冷開水，瀝乾。

02　培根也用滾水燙煮 1 分鐘，撈起、瀝乾。

03　用 1 片培根將尖頭的一半蘆筍捲起（露出筍尖），另一半則再對切，堆放在盤子中間，肉捲圍放在四周。

04　調味汁在小碗中調好，上桌後淋在蘆筍上。

可用醬油膏加糖代替甜醬油，甜醬油為醬油 2 杯加酒 3 湯匙、糖 4 湯匙、八角 1 顆，小火熬煮至濃稠

INGREDIENTS •

15 Asparagus, 15 Bacon slices.

SEASONINGS

½tbsp. Garlic (smashed), 1 ½tbsp. Water, 1 ½tbsp. Sweet soy sauce, 1tsp. Red chili oil, 1tsp. Sesame oil.

STEP BY STEP •

01　Peel asparagus, cut each into halves, blanch in boiling water (add some salt in water), remove and rinse in cold water, drain.

02　Boil bacon for 1 minute, drain.

03　Roll the asparagus (tip part) with bacon slices, (don't cover the tips), Arrange on a plate. Cut the root part into two and put on the center of the plate.

04　Mix the seasonings and pour over the asparagus rolls.

Tips

You may buy the soy sauce paste in the market, add some sugar to instead of sweet soy sauce. To make : put 2 cups of soy sauce, 3 tbsp. wine, 4 tbsp. sugar and 1 star anise, simmer till sticky.

串炸時蔬

Deep-Fried Vegetables Skewers

新鮮香菇、茄子、蕃薯、玉米筍、綠蘆筍、青椒 —————— 隨意

麵粉 ————————————————————————— 3 湯匙

蛋 ——————————————————————————— 1 個

麵包粉 —————————————————————————— 1 杯

步
驟
說
明

01 　將各種蔬菜切成 1 寸長厚塊（香菇對半切），分別用竹籤串起（四種一串）。

02 　蛋打散，加水至⅔杯後，再放麵粉調成稀糊的麵粉糊，麵包粉裝在盤中。

03 　蔬菜串先沾麵粉糊，再輕敷一層麵包粉即可投入七分熱油中，小火慢炸 3 分
　　鐘左右至呈金黃色便可。

Tips

　　上桌時可另附個人喜愛之醬料以供沾食，例如：蘿蔔泥加醬油汁，或蕃茄醬或糖
　　醋汁、A1 牛排醬、辣醬油或美乃滋等。

INGREDIENTS •

Fresh black mushrooms, Eggplants, Baby corns, Sweet potato, Asparagus, each a little, 3tbsp. Flour, 1 Egg, 1cup of Bread crumbs.

STEP BY STEP •

01 　Cut each vegetable into 1" long piece, pierce every four kinds into a skewer.

02 　Beat the egg, add ½~ ⅔ cup of water and flour to make a paste.

03 　Dip vegetable in flour paste and coat with bread crumbs, deep fry in 150℃ oil over low heat for about 3 minutes to golden brown. Remove and serve.

Tips

　A1 sauce, worchester sauce, sweet & sour sauce, ketchup, mayonnaise or soy sauce with smashed turnip can be dipping sauce

干貝扣四蔬

Mold Scallop with Vegetables

使用材料			調味料		
	干貝	3 粒		鹽	1 茶匙
	金菇	1 把		糖	½ 茶匙
	大白菜	1 斤		胡椒粉	少許
	玉米筍	10 支			
	胡蘿蔔	1 小支			
	油	2 湯匙			
	麵粉	2 湯匙			
	高湯（或水）	½ 杯			

01 干貝置碗內加水⅔杯（蓋過干貝），蒸半小時至軟，撕成細條，鋪一層在碗底。

02 金菇切半寸長，用滾水川燙一下撈出，鋪在碗裡干貝上。

03 用熱油 2 湯匙炒大白菜（切條），玉米筍（粗者可對剖為二）和已煮過之胡蘿蔔片，加調味料調味。拌合後瀝出，放在干貝碗中，淋下干貝汁及高湯，入鍋蒸 20 分鐘。

04 取出干貝，泌出湯汁，倒扣在盤中。用油炒香麵粉，再加入湯汁，攪拌成糊狀，淋下少許油拌勻，澆到干貝上供食。

INGREDIENTS •

3 Dried scallops, 100g. Needle mushroom, 600g. Chinese cabbage, 10 Baby corns, 1 Carrot, 2tbsp. Oil, 2tbsp. Flour, ½ cup Soup stock.

SEASONINGS

1tsp. Salt, ½tsp. Sugar, a pinch of Pepper.

STEP BY STEP •

01 Put scallops in a bowl, add ⅔ cup of water, steam for 30 minutes to soft. Drain, tear finely after it cook, put on the bottom of a large bowl.

02 Cut needle mushroom into halves, blanch in boiling water, drain and place on top of scallops.

03 Cut Chinese cabbage into sticks, stir fry with 2tbsp. of heated oil, add baby corn and cooked carrot slices, season with ❶ , stir evenly, drain and stuff in center of the bowl. Pour steamed scallop soup & soup stock, steam for 20 minutes.

04 Pour soup from bowl first, then reverse bowl to let scallop & vegetables stand on a plate. Stir fry oil and flour, add the steamed soup in, mix into a paste, pour over vegetables. Serve.

魚酥翡翠瓜絲

Jade Squash with Crispy Sole

使用材料

扁魚乾	1～2片	糖	¼ 茶匙
絲瓜	1 條	麻油	½ 茶匙
熟胡蘿蔔	半支	胡椒粉、太白粉水	各少許
銀芽	2 兩		
葱花	少許		

調味料

酒	½ 茶匙
鹽	½ 茶匙

01 將扁魚乾放在油中小火煎黃，待涼後切碎。

02 絲瓜刨去外層硬皮後，切成 2 寸長段，再切下外層較綠色部分（約 0.3 公分厚）直切成絲。熟胡蘿蔔也切細絲。銀芽用熱油爆炒一下即瀝出。

03 用 1 湯匙油爆香蔥花，加入胡蘿蔔及絲瓜，大火拌炒，淋下酒，加鹽等調味料調味，加入銀芽拌合，用少許太白粉水使汁濃滑即可裝盤，再撒上扁魚酥即可。

INGREDIENTS •

2 pieces Dried sole, 1 Squash, ½ Carrot cooked, 80g. Mung bean sprouts, Green onion, chopped.

SEASONINGS

❶ ½tsp. Wine, ½tsp. Salt, ¼tsp. Sugar, ½tsp. Sesame oil, a pinch of Pepper, some Cornstarch paste.

STEP BY STEP •

01 Fry sole in warm oil over low heat to golden brown, drain and let stand on tissue, when cooled, chop finely.

02 Peel squash, slice the very outside green part thinly, shred this green part finely. Cut carrot into shreds too. Heat 2tbsp. of oil to stir fry bean sprouts, drain.

03 Heat 1tbsp. of oil to stir fry green onion, add in carrot and squash, stir evenly. Add seasonings ❶, then add bean sprout back, stir fry over high heat. Remove to a plate, sprinkle smashed sole over. Serve.

西拌青花菜

Broccoli Salad with Assorted Dressings

使用材料

| 花椰菜 | 12 兩 |
| 珍珠貝或任何罐裝貝類 | 半罐 |

千島醬

| 美乃滋 | 2 湯匙 |
| 蕃茄醬 | 1 湯匙 |

芥辣醬

| 芥末醬 | 1 湯匙 |
| 美乃滋 | 2 湯匙 |

法式汁

油	3 湯匙
醋	1 湯匙
鹽	½ 茶匙
胡椒粉	少許
沙拉醬	1 湯匙

01　青花菜分小朵後，投入滾水中燙軟，撈出用冷開水沖涼。

02　罐頭貝類切成半寸大小，與青花菜混合後裝盤。

03　法式汁是將調味料（美乃滋除外）放在瓶中，搖晃均勻後再與美乃滋混合均勻。另外千島醬、芥辣醬亦分別調好，裝入小碗中上桌。

Tips

也可挑選一種口味的醬料和青花菜在大碗中拌勻後再上桌。

INGREDIENTS •

450g. Broccoli, ½ Can Clams (any kind).

SEASONINGS

❶ Thousand island:2tbsp. Mayonnaise, 1tbsp. ketchup.

❷ Mustard sauce : 1tbsp. Mustard sauce, 2tbsp. Mayonnaise.

❸ French style : 3tbsp. Oil, 1tbsp. Vinegar, ½tsp. Salt, a pinch of Pepper, 1tbsp. Mayonnaise.

STEP BY STEP •

01　Trim the broccoli into small pieces, blanch in boiling water, then rinse with cold water, drain.

02　Cut clam into ½" cubes, mix with broccoli.

03　Make all dressings respectively, serve as dipping sauce.

Tips

You may choose one kind of dressing to mix with the Broccoli salad.

毛豆八寶醬

Eight Treasure Sauce

01 豬肉丁用少許醬油和太白粉醃過。竹筍和胡蘿蔔煮熟後切丁。毛豆解凍後洗淨（新鮮毛豆用水煮熟）。

02 燒熱油 4 湯匙把肉丁炒熟，盛出。用餘油炒香甜麵醬等調味料，加入肉丁及其他材料，大火拌炒均勻，淋下麻油即可。

INGREDIENTS •

1cup Frozened Soya beans, ½ cup Bamboo shoot, ½cm cubes, ½ cup Dried bean curd, ½cm cubes, ½ cup Pork, ½ cm cubes, 2tbsp. Carrot, ½cm cubes.

SEASONINGS

❶ ½tbsp. Sweet soybean paste, ½tbsp. Hot bean paste, ½tbsp. Soy sauce, 1tsp. Wine, 2tbsp. Water, 1tbsp. Sugar, 1tsp. Sesame oil.

❷ some Soy sauce, some Cornstarch.

STEP BY STEP •

01 Marinate pork with soy sauce and cornstarch. Boil carrot & bamboo shoot (shelled first), cut into ½ cm cubes. Defrozen the soya bean or you may buy the fresh beans to boiled first.

02 Heat 4tbsp. of oil to stir fry pork, drain. Use the remaining oil to stir fry seasonings ❶ and all other ingredients, mix evenly. Serve.

雞茸鮮筍紮

Bamboo Shoot Bundles

01 將筍去殼後煮熟，待冷卻後直剖為四半，分別切薄片，在尖的一邊再加切 3～4 道刀口。

02 雞胸肉絞成肉末，加拌雞料調成雞茸。

03 每 5～6 片筍片為 1 組，攤開後，先撒一層太白粉，再放上雞茸餡，抹平餡後，將筍片捲起成筒狀，用燙軟之韭菜葉紮緊，頂部沾上火腿屑。

04 將雞茸捲放在塗油的碟中，大火蒸 12 分鐘至熟，取出排在盤中。

05 高湯煮滾，調味後用太白粉水勾成稀芡，淋在筍絮上即可。

INGREDIENTS •

3 Bamboo shoots, 150g. Chicken breast, 1tbsp. Ham, chopped, a few stalks Leek, 1cup Soup stock.

SEASONINGS

❶ ¼tsp. Salt, 1tsp. Wine, 1tbsp. Egg white, 1tsp. Cornstarch.

❷ ¼tsp. Salt, a pinch of Pepper, some Cornstarch paste.

STEP BY STEP •

01 Boil the bamboo shoot, remove shell, let cool. Cut along the grain into 4 parts, then cut into very thin slices. On ⅔ part of the tip, cut into thin shreds like a comb.

02 Smash chicken breast, mix with ❶.

03 Connect every 5~6 pieces of bamboo shoot together into a flat sheet, sprinkle over cornstarch, spread chicken paste, roll the bamboo shoot into a roll, fasten with a blanched leek. Dip the end with chopped ham.

04 Rub some oil on a plate, put the rolls on, steam over high heat for 12 minutes.

05 Cook the soup stock, season with salt, thicken with cornstarch paste. Bring to a boil, pour over rolls. Serve.

碧綠三色菇

Tri-Color Mushrooms

01 青江菜摘菜心後在熱水中（加少許鹽）燙熟，排在盤中呈三角放射型。

02 洋菇去蒂洗淨，用 1 湯匙油爆香蔥段，放下洋菇炒香，淋高湯並調味煮 3 分鐘，勾芡加奶水調勻，排在盤中⅓處。

03 另用 1 湯匙油爆香蔥段，放下草菇炒香，加草菇料煮 1 分鐘即可勾芡裝盤。

04 用 1 湯匙油炒蔥絲及金菇，加蠔油及高湯煮片刻，撒下青椒、紅辣椒絲並調味，勾芡裝入盤中。

INGREDIENTS ·

12 pieces Mushroom, 12 pieces Straw mushroom, 240g. Needle mushroom, a few of Green cabbage, 1 stalk Green onion.

SEASONINGS

❶ 6 pieces Green onion sections, 1cup Soup stock, ½tsp. Salt, a little of Sugar, some Cornstarch paste, 2tbsp. milk.

❷ 6 pieces Green onion section, ⅔ cup Soup stock, 1 ½tbsp. dark colored Soy sauce, ½tsp. Sugar, some Cornstarch paste.

❸ a little of Green onion shreds, a little of Green pepper shreds, a little of Red chili shreds, ½tbsp. Oyster sauce, ¼ cup Soup stock, some Cornstarch paste.

STEP BY STEP ·

01 Trim the green cabbage and blanch in boiling water, arrange in a Benz's-car-sign design to divide into three spaces for three kinds of mushrooms.

02 Heat 1tbsp. of oil to stir fry green onion, add mushroom and ❶. Cook for 3 minutes. Place on the plate.

03 Heat 1tbsp. of oil to stir fry green onion, add straw mushroom and ❷, cook for 1 minute. Place on the plate.

04 Use the same way to stir fry needle mushroom and ❸. Place in the plate.

魚香溜藕夾

Lotus Root's Cake—Szechuan Style

使用材料

鮮藕	2 節	
絞肉	4 兩	
蔥花	1 湯匙	

拌肉料

蔥、薑屑	各 1 茶匙	
醬油、水	各 ½ 湯匙	
酒	1 茶匙	
太白粉	1 茶匙	

胡椒粉	¼ 茶匙	

魚香料

蔥、薑、蒜末	各 1 茶匙	
辣豆瓣醬	1 湯匙	
醬油、糖、醋	各 1 茶匙	
鹽	¼ 茶匙	
麻油、太白粉	各 ½ 茶匙	
水	4 湯匙	

步驟說明

01　絞肉加入拌肉料拌勻，調成肉餡。

02　鮮藕削去外皮，切成薄片，平擺在菜板上，撒一層太白粉，每 2 片中夾上絞肉餡，做成藕夾。

03　麵粉加水調成麵粉糊，藕夾沾裹麵糊後，用熱油炸熟。

04　用 1 湯匙油炒香魚香料，撒下蔥花，放下藕夾一拌即可裝盤。

Tips

四川菜中有名的辣豆瓣魚，因所用的辛香料與調味料眾多，所形成的味道很好，於是就將這個味道來燒其他的食材，並冠以「魚香」之名，例如：魚香肉絲、魚香茄子等菜式。

INGREDIENTS •

2 section Lotus roots, 150g. Minced pork, 1tbsp. Green onion, chopped.

SEASONINGS

❶ Green onion & Ginger, chopped, each 1tsp., ½tbsp. Soy sauce, ½tbsp. Water, 1tsp. Wine, 1tsp. Cornstarch, ¼tsp. Pepper.

❷ Chopped ginger, chopped green onion, chopped garlic each 1tsp., 1tbsp. Hot bean paste, 1tsp. Soy sauce, 1tsp. Sugar, 1tbsp. Vinegar, ¼tsp. Salt, ½tsp. Sesame oil, ½tsp. Cornstarch, 4tbsp. Water.

STEP BY STEP •

01　Mix minced pork with ❶ , stir to very sticky.

02　Peel lotus roots, cut into thin slices against the grain, arrange on a plate, sprinkle cornstarch over, put the pork between every two lotus root slices like burger shape.

03　Mix flour with water to make batter, use this to coat lotus root burger and deep fry in heated oil to done, drain.

04　Heat 1tbsp. of oil to cook ❷ , add green onion, pour lotus root in, mix and remove it to a serving plate. Serve.

Tips

Sichuan cuisine is famous for its spicy fish in thick bean sauce. It tastes good because of the many spices and seasonings used; so I use the flavor from this sauce, named "fish sauce," to make fish-flavored shredded port, fish-flavor eggplant and other such dishes.

什錦蔬菜沙拉

Assorted Vegetable Salad

使用材料

大顆馬鈴薯	1 個
冷凍什錦蔬菜	2 杯
培根	3 片
生菜葉	數張

調味料

美乃滋	2 湯匙
鹽	½ 茶匙
胡椒粉	少許

01 馬鈴薯煮熟去皮，切成 1 公分四方丁。冷凍蔬菜用滾水燙煮一下，隨即撈出。培根用小火煎黃，待冷切小丁。

02 上項材料放大碗中加調味料拌勻，裝入圓碗中，移入冰箱中略冰過，食前扣在大盤中，撒下培根小丁即可，食時可用生菜或麵包包捲食用。

INGREDIENTS •

400g. Potato, 2cups Frozen vegetables(Assorted), 3 slices. Bacon, Lettuce leaves(several pieces).

SEASONINGS

2tbsp. Mayonnaise, ½tsp. Salt, a pinch of Pepper.

STEP BY STEP •

01 Boil potato, peel off skin, cut into 1" cubes. Blanch the frozen vegetables in boiling water, drain. Fry the bacon to browned, cut into small cubes.

02 Mix potato, vegetables & seasonings into a large mold, freeze for a while, then reverse up-side-down to a plate, sprinkle the bacon. Serve and wrapped with lettuce leaves.

什錦蔬菜沙拉捲

Vegetable Salad Rolls

01　什錦蔬菜沙拉中加入洋火腿丁及鳳梨丁拌勻。（什錦蔬菜沙拉做法請參考 P.154。）

02　豆腐衣切成 3 小張，包入沙拉，小火熱油炸至外皮酥黃即可瀝出。（用春捲皮包捲可在炸過後，再一切為二裝盤。）

INGREDIENTS •

2 cups Assorted Vegetables salad, 2 pieces Ham slice, 2 pieces Pineapple slice, 5 pieces egg roll wrapper (or dried bean curd sheet).

STEP BY STEP •

01　Mix assorted vegetable salad with ham cubes & pineaple cubes. (refer to p.154 for vegetable salad.)

02　Cut each bean curd sheet into 3 pieces, wrap in salad mixture, fold into a roll, deep fry in heated oil to golden brown, Serve. (Or wrap it with egg roll wrapper then deep fry it.)

魚香汁拌四季豆

Green Beans with Szechuan Sauce

使用材料

牛肉或豬肉	3 兩	水	½ 杯
四季豆	半斤	鹽、糖	各少許
胡蘿蔔絲	⅓ 杯	太白粉	1 茶匙
蔥花	1 湯匙	麻油	1 茶匙

魚香料

薑、蒜屑	各 ½ 茶匙
辣豆瓣醬	1 湯匙
酒、醬油	各 ½ 湯匙

01　牛肉或豬肉煮熟切絲，胡蘿蔔切成絲，用少許鹽抓醃過；四季豆摘除老筋，煮熟、瀝出、吹涼，全部放在大碗中。

02　用 1 湯匙油爆香薑、蒜屑及辣豆瓣醬，加入其他調勻的魚香料，撒下蔥花，淋入四季豆中，拌勻即可。

INGREDIENTS •

120g. Pork, 300g. Green beans, ⅓ cup Carrot shreds, 1tbsp. Green onion, chopped.

SEASONINGS

½tsp. Garlic, chopped, ½tsp. Ginger, chopped,1tbsp. Hot bean paste, ½tbsp. Wine, ½tbsp. Soy sauce, ½ cup Water, a pinch of Salt, a pinch of Sugar, 1tsp. Cornstarch, 1tsp. Sesame oil.

STEP BY STEP •

01　Boil pork and cut into strings. Marinate carrot shreds with a little of salt. Blanch green bean to done, drain and let cool, put all ingredients in a large bowl.

02　Heat 1tbsp. of oil to stir fry ginger and garlic, add seasonings, bring to a boil, sprinkle green onion, mix with green beans. Serve.

肉燥灼拌四季豆

Green Beans with Minced Pork Sauce

01　四季豆摘除老筋，太長的可一切為二。投入滾水中川燙至軟，瀝出、放入大碗中。冷凍四季豆僅需一切為二之後在熱水中灼燙 10 秒鐘。

02　將熱的肉燥及醬油膏淋入碗中，撒下蔥花，拌勻即可裝盤上桌。

Tips

肉燥做法參考「香菇肉燥蒸鮭魚」的做法 P.100。

INGREDIENTS •

300g. Green beans, 2tbsp. Minced pork sauce, 1tbsp. Soy sauce paste, 1tbsp. Green onion, chopped.

STEP BY STEP •

01　Remove grain ends from green beans, cut each into 2 parts, blanch in boiling water, drain and put into a large bowl. If you use the frozen green beans, you may just halve the green beans and blanch for 10 seconds.

02　Add heated pork sauce, soy sauce paste, and green onion in, mix well with green beans. Serve.

Tips

Refer the procedures of pork sauce in page 100.

豌豆濃湯

Snow Peas Cream Soup

使用材料		
冷凍豌豆		2 杯
洋蔥屑		2 湯匙
清湯		3 ½ 杯
植物性奶油		3 湯匙
麵粉		3 湯匙
鹽		1 茶匙
牛奶		⅓ 杯
麵包丁		2 湯匙

01　將冷凍豌豆解凍後，放入果汁機中，加入清湯（或水）2 ½ 杯打成糊狀，過濾 1 次。

02　炒鍋中將奶油先溶化，放下洋蔥屑炒香，再加入麵粉炒黃，將清湯 1 杯慢慢淋下，同時用鏟子將麵糊攪散至均勻，再加入豌豆糊調勻。

03　煮滾後將洋蔥屑撈棄，放鹽調味，加入牛奶即可裝入個人用湯碟或杯子中，上桌後撒下炸黃之脆麵包丁即可（亦可撒炸酥之培根丁或洋火腿絲）。

INGREDIENTS •

2 cups Frozen snow peas, 2tbsp. Onion, chopped, 3 ½cups Soup stock, 3tbsp. Butter, 3tbsp. Flour, 1tsp. Salt, ⅓ cup Milk, 2tbsp. toast cubes.

STEP BY STEP •

01　Defroze the snow peas, put into a blender, add 2 ½cups of soup stock, blend finely, drain.

02　Heat the butter, stir fry onion, add flour, then add 1cup of soup stock gradually, stir the flour paste evenly, add in snow pea juice.

03　Bring to a boil, sieve the onion off, season the soup with salt, add milk. Serve with deep-fried toast cubes (or fried bacon).

炒豌豆雞絲

Chicken Strings with Snow Peas

01 雞胸肉去皮除筋後，直紋切成細絲，用醃雞料仔細拌勻，醃 10 分鐘以上。

02 冷凍豌豆投入滾水中燙一下即刻瀝出。

03 鍋子燒熱後倒下油 1 ½ 杯，待七分熱時，放下雞絲用筷子快速將雞絲撥散，待雞絲變白夠熟時便盛出。

04 另用 1 湯匙油爆香蔥屑後，將豌豆粒及雞絲一起下鍋，淋下酒及鹽，拌炒均勻後用太白粉水勾芡，再淋入數滴麻油即可。

INGREDIENTS •

½ Chicken breast, 1 ½cups Frozen snow peas, 1tbsp. Green onion, chopped.

SEASONINGS

❶ 1tbsp. Egg white. ¼tsp. Salt, ½tsp. Cornstarch.

❷ 1tsp. Wine, ½tsp. Salt, ½tbsp. Cornstarch paste, a few drops of Sesame oil.

STEP BY STEP •

01 Remove skin from chicken, cut into fine strings along the grain, marinate with ❶ for 10 minutes.

02 Blanch the snow peas, drain.

03 Heat 1 ½cups of oil to 140°C, stir fry chicken shreds, separate with chopsticks. When chicken turns white, drain.

04 Heat 1tbsp. of oil, stir fry green onion, add chicken & snow peas, add seasonings ❷ , mix well. Serve.

饊子素菜捲

Crispy Vegetarian Rolls

步驟說明

01 春捲皮塗上美乃滋，放生菜葉 1 枚，再塗美乃滋少許，再放上 1 支油炸饊子，撒一些素肉鬆或芝麻或花生粉，捲成筒狀。

02 將頭尾略切除一些，再對切為二即可。

▸ 可用吃烤鴨用的單餅代替春捲皮。

▸ 也可以吃葷的，內捲叉燒肉絲、蝦仁、醬肉、肉鬆或加包蘆筍、豆腐乾等。

INGREDIENTS •

4 San-Tsu (Deep Friend Twister Bar), 4 pieces Lettuce leaf, 4tbsp. Vegetarian meat fluff, 4 pieces egg roll wrapper, Mayonnaise.

STEP BY STEP •

01 Rub mayonnaise on the egg roll wrapper, place a piece of lettuce leaf, put 1 San-Tsu on, sprinkle the vegetarian meat fluff, fold and roll into a cylinder.

02 Cut off two ends and cut each roll into halves.

· You may use the flour pan sheet (for Pei-jin Duck) to substitute for the egg roll wrapper.

· You may change the vegetarian ingredients to assorted meats, such as Bar-B-Q pork, shrimp, meat sauce or asparagus or dried bean curd.

RECOMMENDED BOOKS
好書推薦

書 名	五味八珍的歲月
作 者	傅培梅
定 價	350 元
ISBN	978-986-950-170-5

烹飪教育家傅培梅，用美食溫暖了每個台灣人的胃，且看一個連餃子都煮不好的家庭主婦，如何變身名聞國際的烹飪大師？酸鹹苦辣甜，細看傅培梅老師以人生調味出的五味八珍的歲月。

書 名	傅培梅的中國名菜精選（精裝）
作 者	傅培梅
定 價	1500 元
ISBN	978-986-364-024-0

全書包含廣東、台灣、福建、四川、江浙、湖南、北京等區域 251 道經典佳餚，均是各地區家喻戶曉，深受大家喜愛的家常菜和小菜等在內。

書 名	傅培梅時間的美味中國菜（精裝）
作 者	傅培梅，程安琪
定 價	650 元
ISBN	978-986-935-910-8

傅培梅對中國烹調藝術的詮釋，700 道蔬菜、肉類、海鮮、菇類、豆製品、蛋類等食譜，詳述連主廚都一知半解的火候、刀工、烹調技法。

書 名	培梅家常菜（書 +DVD）
作 者	傅培梅，程安琪
定 價	450 元
ISBN	978-986-917-162-5

國寶級烹飪大師傅培梅老師示範解說，從選購食材到火候刀工，最食用的烹調法則一次學會，是家家必備的廚房寶典，中英對照，人人可學。

書　名	媽媽的菜：傅培梅家傳幸福的滋味
作　者	程安琪
定　價	380 元
ISBN	9789869797214

本書集結傅培梅老師所出版過的食譜，重新編排，並透過程安琪老師一一為讀者解說，一共五大章、82 道食譜的動人故事。希望大家也能嚐到這些幸福的味道！

書　名	美味台灣菜：138 道傳統美味與流行創業小吃
作　者	傅培梅，程安琪，陳盈舟
定　價	548 元
ISBN	978-986-364-125-4

一次收集 79 道傳統名菜 X 59 種台灣小吃，開店創業、在家品味不可缺少的美食寶典。傅培梅、程安琪、陳盈舟三位老師傳承復古好味道，細心講解獨家好吃秘方，讓您做出道地台灣美味。

書　名	外傭學做銀髮族餐點（中菲印對照）
作　者	程安琪
定　價	300 元
ISBN	978-986-364-150-6

精選 60 道料理做法搭配中、菲、印 3 種語言對照，大火快炒、小火慢燉，清蒸涼拌、燒滷烘烤……主菜、小菜、飯粥到湯品，教外籍朋友輕鬆學做菜。

書　名	無油煙烤箱菜輕鬆做：只需一台烤箱，海鮮肉類 X 主食 X 點心豐富上桌！
作　者	程安琪
定　價	300 元
ISBN	978-986-069-219-8

「烤」是很簡單的烹調法，溫控、定時、無油煙，且失敗率低，烤好之後自然散發的出的香氣，非常吸引人，不只烤雞、肉串及烘培點心，就連年輕人最喜愛的流行焗烤菜，都能輕鬆端上桌。

五味八珍的餐桌
品牌故事

60 年前，傅培梅老師在電視上，示範著一道道的美食，引領著全台的家庭主婦們，第二天就能在自己家的餐桌上，端出能滿足全家人味蕾的一餐，可以說是那個時代，很多人對「家」的記憶，對自己「母親味道」的記憶。

程安琪老師，傳承了母親對烹飪教學的熱忱，年近 70 的她，仍然為滿足學生們對照顧家人胃口與讓小孩吃得好的心願，幾乎每天都忙於教學，跟大家分享她的烹飪心得與技巧。

安琪老師認為：烹飪技巧與味道，在烹飪上同樣重要，加上現代人生活忙碌，能花在廚房裡的時間不是很穩定與充分，為了能幫助每個人，都能在短時間端出同時具備美味與健康的食物，從 2020 年起，安琪老師開始投入研發冷凍食品。

也由於現在冷凍科技的發達，能將食物的營養、口感完全保存起來，而且在不用添加任何化學元素情況下，即可將食物保存長達一年，都不會有任何質變，「急速冷凍」可以說是最理想的食物保存方式。

在歷經兩年的時間裡，我們陸續推出了可以用來做菜，也可以簡單拌麵的「鮮拌醬料包」、同時也推出幾種「成菜」，解凍後簡單加熱就可以上桌食用。

我們也嘗試挑選一些熟悉的老店，跟老闆溝通理念，並跟他們一起將一些有特色的菜，製成冷凍食品，方便大家在家裡即可吃到「名店名菜」。

傳遞美味、選材惟好、注重健康，是我們進入食品產業的初心，也是我們的信念。

冷凍醬料做美食

程安琪老師研發的冷凍調理包，讓您在家也能輕鬆做出營養美味的料理。

冷凍醬料的 5 大優點

省調味 × 超方便 × 輕鬆煮 × 多樣化 × 營養好

選用國產天麴豬，符合潔淨標章認證要求，我們在材料和製程方面皆嚴格把關，保證提供令大眾安心的食品。

三友官網

五味八珍的
餐桌官網

五味八珍的
餐桌 FB

程安琪
鮮拌味 FB

程安琪入廚
40 年 FB

五味八珍的
餐桌 LINE @

聯繫客服 電話：02-23771163 傳真：02-23771213

冷凍醬料調理包

香菇蕃茄紹子

歷經數小時小火慢熬蕃茄，搭配香菇、洋蔥、豬絞肉，最後拌炒獨家私房蘿蔔乾，堆疊出層層的香氣，讓每一口都衝擊著味蕾。

雪菜肉末

台菜不能少的雪裡紅拌炒豬絞肉，全雞熬煮的雞湯是精華更是秘訣所在，經典又道地的清爽口感，叫人嘗過後欲罷不能。

麻辣紹子

麻與辣的結合，香辣過癮又銷魂，採用頂級大紅袍花椒，搭配多種獨家秘製辣椒配方，雙重美味、一次滿足。

北方炸醬

堅持傳承好味道，鹹甜濃郁的醬香，口口紮實、色澤鮮亮、香氣十足，多種料理皆可加入拌炒，迴盪在舌尖上的味蕾，留香久久。

冷凍家常菜

一品金華雞湯

使用金華火腿（台灣）、豬骨、雞骨熬煮八小時打底的豐富膠質湯頭，再用豬腳、土雞燜燉 2 小時，並加入干貝提升料理的鮮甜與層次。

靠福·烤麩

一道素食者可食的家常菜，木耳號稱血管清道夫，花菇為菌中之王，綠竹筍含有豐富的纖維質。此菜為一道冷菜，亦可微溫食用。

3 種快速解凍法

想吃熱騰騰的餐點，就是這麼簡單

1. 回鍋解凍法
將醬料倒入鍋中，用小火加熱至香氣溢出即可。

2. 熱水加熱法
將冷凍調理包放入熱水中，約 2 ～ 3 分鐘即可解凍。

3. 常溫解凍法
將冷凍調理包放入常溫水中，約 5 ～ 6 分鐘即可解凍。

私房菜

純手工製作，交期較久，如有需要請聯繫客服
02-23771163

程家大肉

紅燒獅子頭

頂級干貝 XO 醬

創意家常菜

CREATIVE CHINESE HOME DISHES

書　　　名	創意家常菜
作　　　者	傅培梅
修 訂 版	程安琪
校　　　對	
主　　　編	莊旻嬪
美　　　編	譽緻國際美學企業社
封 面 設 計	譽緻國際美學企業社‧羅光宇
攝 影 師	陳弘暐
發 行 人	程安琪
總 編 輯	盧美娜
發 行 部	侯莉莉
財 務 部	許麗娟
印　　　務	許丁財
法 律 顧 問	樸泰國際法律事務所許家華律師
藝 文 空 間	三友藝文複合空間
地　　　址	106 台北市安和路 2 段 213 號 9 樓
電　　　話	（02）2377-1163
出 版 者	橘子文化事業有限公司
總 代 理	三友圖書有限公司
地　　　址	106 台北市安和路 2 段 213 號 4 樓
電　　　話	（02）2377-4155
傳　　　真	（02）2377-4355
E-mail	service@sanyau.com.tw
郵 政 劃 撥	05844889 三友圖書有限公司

總 經 銷	大和書報圖書股份有限公司
地　　　址	新北市新莊區五工五路 2 號
電　　　話	（02）8990-2588
傳　　　真	（02）2299-7900
初　　　版	2022 年 7 月
定　　　價	新臺幣 380 元
I S B N	978-986-364-192-6（平裝）

國家圖書館出版品預行編目（CIP）資料

創意家常菜 / 傅培梅作. -- 初版. -- 臺北市：橘子
文化事業有限公司, 2022.07
　　面；　公分
　　ISBN 978-986-364-192-6（平裝）

1.CST: 食譜

427.1　　　　　　　　　　　　　111007432

三友官網　　三友 Line@